Texts in Philosophy

Volume 37

Introduction to the Philosophy of Probability

The Different Interpretations

Volume 27
Unorthodox Analytic Philosophy
Guillermo E. Rosado Haddock

Volume 28
Quantum Heresies.
Kent A. Peacock, with a foreword by James Robert Brown

Volume 29
Carrollian Notes
George Englebretsen

Volume 30
Problems from Hume
David Bostock

Volume 31
The Labyrinth of Infinity or the Enigma of Existence
Stathis Livadas

Volume 32
Experience and the Ultimate Structure of Reality, Husserl's Pursuit of Truth
Claire Ortiz Hill

Volume 33
Analytic Philosophy: A User's Guide
Andrew David Irvine

Volume 34
Doing Logic. The Continuum through Sets
Montgomery Link

Volume 35
World 3. Ontological and Epistemological Studies
Ilkka Niiniluoto

Volume 36
Georg Cantor's Metaphysical Correspondence
Claire Hill and Walter Tydecks

Volume 37
Introduction to the Philosophy of Probability. The Different Interpretations
Willian Silva Barros and Carlos Alberto Miraglia

Texts in Philosophy Series Editors
Vincent F. Hendriks vincent@hum.ku.dk
John Symons jsymons@utep.edu
Dov Gabbay dov.gabbay@kcl.ac.uk

Introduction to the Philosophy of Probability
The Different Interpretations

Willian Silva Barros
Carlos Alberto Miraglia

ISBN 978-1-84890-524-5

College Publications
Scientific Director: Dov Gabbay
Managing Director: Jane Spurr

http://www.collegepublications.co.uk

Image on front cover designed by Laura Pereira Barros
Original cover design by Laraine Welch

DEDICATION

To my wife, Giselda Maria Pereira, my love, my safe harbour and my life.

To my daughters:

Luana Fonseca Barros, my Moon who brings calm;
Laura Pereira Barros, my Star who guides;
Aurora Brum Barros, my Sun who warms.

Willian Silva Barros

Contents

Preface

When a meteorologist announces a '70% probability of rain', the same word seems to circulate quite naturally, used by a physicist to describe the decay of a particle or by a statistician when calculating a p-value, as though 'probability' had a uniform meaning. Yet the concept at stake is not always the same, and this apparent uniformity is a recurrent source of scientific and philosophical misunderstanding. This book begins with that conceptual impasse.

Probability is widely used in modern science as a means of quantifying uncertainty and, paradoxically, is one of the concepts whose meaning is least often discussed. We learn to calculate it from an early stage and it is employed in statistics, physics, agronomy, biology, economics and the social sciences; yet comparatively little is said about exactly what may be asserted when a probability is assigned.

We speak of real possibility, frequency, degree of belief, evidential confirmation and causal tendency; it is often assumed that these expressions all point to a single thing. The aim here is to unsettle that naturalised common sense by making explicit the conceptual cost of each choice in terms of ontological and epistemic commitments, limits of application and assumptions about rationality, while also making explicit the promise that each interpretation makes to the scientist and the philosopher.

In Portuguese there are works in the scientific literature on mathematical probability and statistics, as well as important philosophical discussions of science, rationality and method. Even so, works in the same language that take the philosophy of probability as their central subject, with a broad historical scope extending from the emergence of modern probability calculus in the seventeenth century to contemporary formulations, are less common. Works offering a systematic comparison of competing interpretations are also less common, particularly those that neither treat the subject merely as methodological justification in an introductory chapter of a technical manual nor reduce it to a formalism detached from its epistemological consequences, as happens when the apparatus of theorems is privileged without discussion of what probability assignments mean.

We wrote this book to help fill that gap. Its foundation was Willian Silva Barros's undergraduate dissertation in Philosophy, developed under the supervision of Professor Carlos Alberto Miraglia. The present volume is a joint work by both authors. We also wrote it to offer the reader a 'map': not a catalogue of schools, but an instrument for recognising, by explicit criteria, what is at stake when one interpretation of probability is chosen over another, even when that choice presents itself as no more than a disciplinary convention.

The point of departure is a duality that has accompanied probability since its modern consolidation. On the one hand, probability is understood as something attributed to the world: as objective chance or, in different versions, as objective regularity, stable pattern or causal disposition. On the other hand, it is understood as something relative to knowledge: as a degree of rational belief, a measure of evidential confirmation and a tool for guiding rational decision under uncertainty.

This tension is not merely a matter of vocabulary. It structures what each tradition understands by objectivity. For some, to be objective is to be independent of the agent's epistemic states; hence the appeal of frequentist readings and, in certain versions, propensity readings. For others, objectivity consists in rational determination, under explicit rules, given the evidence; hence projects in inductive logic such as that of Rudolf Carnap (1891–1970). For still others, the normative core lies in the coherence of beliefs and bets in the technical sense; hence the line of thought running through Frank Ramsey (1903–1930) and Bruno de Finetti (1906–1985).

This axis is heuristic rather than a rigid binary. Some contemporary approaches seek intermediate or hybrid positions, as do certain versions of objective Bayesianism.

In many areas of physics, ordinary talk of 'objective chances' tends to point towards the ontological pole. In many applications of statistics and decision theory, the practice of updating beliefs in the light of evidence tends to point towards the epistemic pole. This book treats that axis as its central conceptual bridge for comparison, returning to it in every chapter precisely because this is where the divergence takes practical form and ceases to look like a mere dispute between schools.

There is also a contemporary reason to return seriously to the question of what probability means. In a world increasingly mediated by algorithms, data science and decisions under uncertainty, from medicine and epidemiology to economics and public policy, probability is no longer merely a technical tool. It has become a language of justification: a language of risk, evidence and public trust.

Understanding its philosophical foundations is not an academic ornament. It is a condition for using responsibly a concept that organises real choices and produces public effects, as was evident, for example, in the communication of risk and evidence during public-health crises, when interpretative decisions directly influenced trust and adherence to policy.

When probability is treated uncritically, it tends to conceal modelling and interpretative choices as though they followed automatically from the formalism. When it is examined by explicit criteria, its use becomes more honest, more transparent and, ultimately, more rigorous.

The book is organised to lead the reader from the problem to the conceptual map and then from the map to comparative conceptual and methodological tests. Chapters 1 and 2 reconstruct the historical and conceptual formation of probability, from games of chance to the development of statistics and the entry of probability into physics as an explanatory resource. These chapters also make explicit the role of modern axiomatic formalism and the separation between the rules of the calculus and interpretative commitments.

Chapters 3 to 7 present the classical (equiprobability), frequentist, propensity, inductive-logical and subjective interpretations with symmetrical treatment, so that each is assessed by its performance when faced with recurrent problems in the interpretation of probability rather than by historical prestige or personal preference. Within the discussion of objective chance, the contemporary best-system approach is also examined; it reduces neither to frequentism nor to propensity theory.

Chapters 8 to 10 then place the interpretations in direct dialogue through cross-cutting debates and case studies, showing where each position is strong, where it fails and which conceptual assumptions it tends to adopt when confronted with its limits. The aim is to equip readers to recognise what is being asserted in their own field whenever a probability is assigned.

The criteria of comparison are deliberately explicit. We ask whether an interpretation can deal with singular, non-repeatable events without resorting to ad hoc auxiliary hypotheses; whether it depends on infinite sequences and mathematical idealisations; how it confronts the reference-class problem, that is, the selection of relevant cases; how it understands probabilistic conditioning and the updating of beliefs; what relation it establishes between probability and decision under uncertainty; and, above all, what it presupposes when assigning initial probabilities.[1]

Each chapter seeks to show both what the interpretation explains well and what it assumes as a conceptual cost in order to explain it.

In the end, the aim will not be victory by exhaustion. It will be to provide the reader with a criterion-guided synthesis capable of informing conscious interpretative choices: a final map that, by honestly acknowledging the costs of each route, makes it possible to navigate the conceptual impasse with which the book began.

[1]Ad hoc auxiliary hypotheses are additional assumptions introduced solely to save a hypothesis from a counterexample, without independent justification and without generating new testable predictions.

Introduction: the problem of the term

(The semantic-epistemic problem of probability)

Probability and philosophical analysis

Some concepts enter science through the door of technique and eventually demand philosophical analysis and a metaphysical position. Probability is one of them.

When Blaise Pascal (1623–1662) and Pierre de Fermat (1601–1665) exchanged letters in 1654 about problems of games of chance, a calculus was consolidated and a conceptual tension emerged. How is uncertainty to be formalised mathematically?

In everyday use, probability appears simple. Something is said to be 'probable' when it is neither certain nor impossible. In a more disciplined register, a number between 0 and 1 is assigned to an event as a measure of its chance, plausibility or confirmatory strength in the light of evidence. That description is harmless in itself. The difficulty begins when we ask what, exactly, that number is intended to represent in the context in which it is used.

If probability were merely an instrument of calculation, it would be enough to learn the rules of combination, conditioning and independence and then apply them. But probability does not remain confined to calculation. It guides real decisions, supports inferences, justifies policies, warrants predictions and structures scientific explanations.

A doctor estimates the risk associated with a treatment, a researcher assesses the strength of evidence, an insurer prices a contract, a physicist describes radioactive decay and a data scientist classifies an individual as 'high risk'. In all these cases, a probabilistic number enters as a premise in reasoning that purports to be rational and, often, objective.

Philosophy becomes necessary because the question 'what is probability?' is not merely a matter of academic conceptual precision. It structures the very logic of scientific and practical justification by determining what counts as evidence, what counts as objectivity and what counts as error.

Probability demands philosophy because, unlike operational quantities such as mass or temperature, whose physical meaning tends to stabilise around a predominantly objective referent, although not one immune from philosophical discussion, probability seems to oscillate between two domains.

In some contexts, we speak as though probability were a feature of the world. In others, we speak as though it were a feature of our knowledge. The same person may alternate between these readings in the same discourse without noticing, as though they were merely different ways of speaking.

The introduction to this book proposes to interrupt that unconscious alternation and make it a subject of enquiry, for it is precisely there that the interpretations diverge.

The two faces of probability: objective chance and epistemic state

The duality can be stated directly. There is an ontological reading in which probabilities are chances, tendencies, propensities, objective regularities or constitutive elements of physical processes. On this reading, which concerns the nature of reality, probability refers to the world, or at least to certain domains of it, such as some mass phenomena, stochastic processes and, for many, quantum events.

There is also an epistemic reading in which probabilities express degrees of rational belief, levels of confirmation, measures of the extent to which a hypothesis is supported by evidence, or normative rules for decision under uncertainty. On this reading, probability refers to an agent's epistemic state, understood as a personal and subjective degree of belief; or to an objective logical relation of confirmation between hypotheses and evidence; or to an informational state capable of normative regulation.

This twofold character echoes broader philosophical divisions. Realism and anti-realism, objectivism and subjectivism, epistemic externalism and internalism reappear here in a disciplined form, because what is at stake is not merely a theoretical formality. It is the question of the status of the probabilistic number. Is it intended to name something in the world, as a feature of reality, or to order our access to the world, as a norm of belief and decision?

The difference is not merely one of ‘general philosophy’. It changes the regime of intelligibility of the same statement: what counts as justification, what counts as error and even what would count as refutation. Consider the statement ‘the probability of rain tomorrow is 70%’. At first sight, the sentence appears transparent, as though the number were simply a quantitative portrait of a fact.

Yet the same sentence may function as a summary of frequencies within reference classes; as a statement about physical chance given an atmospheric configuration; as an expression of an agent's degree of rational belief in the light of data and models; or merely as an operational value for guiding a decision under loss and risk. The wording remains the same, but its content does not. The conditions under which it would be true change, the evidential criteria supporting it change and the practical implications it licenses change. In other words, the interpretative commitment changes and, with it, the meaning and scope of what is being asserted.

The same conceptual duality reappears within science itself, not as a terminological dispute but as a disagreement about the status of a probabilistic statement. In applied statistics, when one speaks of the probability that a parameter lies within an interval, many frequentists will say that, strictly speaking, this form of statement should be avoided, because parameters are not random events and probabilities concern repeated procedures and observed frequencies. A Bayesian, by contrast, will say that this is precisely what one wishes to express: not a frequency in the world, but a degree of rational belief conditioned on data and a model.

In quantum mechanics the tension becomes especially acute. The wave function encodes probabilities, but probabilities of what? Of measurement outcomes, as though they were chances in the world; or of our knowledge of systems, as maintained by interpretations according to which quantum probability expresses a fundamental epistemic constraint on the agent?

This book takes the twofold character of probability as an axis, not in order to reduce everything to two labels, but to guide the reader through the comparison.

For each interpretation, we ask which part of the puzzle it aims to solve and what price it pays for doing so. Some positions secure objectivity at the cost of restricting their domain of application. Others secure generality at the cost of accepting an unavoidable element of judgement. Some preserve the vocabulary of 'real chances'. Others preserve the normativity of rational belief.

The conflict is not accidental. It is the very landscape of the problem.

Axiomatic formalism and meaning

The mathematical modernity of probability has a widely recognised landmark. Before Andrei Kolmogorov (1903–1987), probability was a mathematical concept with variable and, at times, inconsistent foundations. In 1933, Kolmogorov's axiomatic formulation integrated probability calculus into measure theory.

The impact of this move was not merely technical. It made explicit a distinction that is philosophically decisive. Kolmogorov's axiomatisation defines the rules of the calculus, but does not by itself define the meaning of what is calculated.

Formally, the axiomatisation begins with a sample space of possible outcomes, a structure of events and a function assigning to each event a number between 0 and 1, subject to three basic axioms.

The non-negativity axiom rules out probabilities below zero. The normalisation axiom stipulates that the certain event has probability 1. The additivity axiom ensures that, for mutually exclusive events, the probability of their union is the sum of their probabilities.[2]

These three principles define the rules of the mathematical game, but they do not tell us which game is being played when those rules are applied to the world.

This construction is powerful. It makes it possible to deal, within a single formal language, with coins, urns, continuous processes, random variables, distributions, independence and conditioning. It also makes it possible to prove theorems explaining why frequencies tend to stabilise and why sums of many influences tend to behave approximately normally.

None of this, however, determines what 'probability' is in the philosophical sense. The formalism is compatible with very different readings. It may be interpreted as a limiting frequency in an ideal sequence, as a coherent degree of belief, as a measure of confirmation or as an expression of physical chance. The axioms do not choose. They organise. They guarantee the internal consistency of the calculus, but impose neither an ontology nor an epistemology.

[2]A certain event means that the normalisation axiom fixes $P(\Omega) = 1$, that is, the universal event of the model.

This semantic underdetermination, or interpretative neutrality of the formalism, is a mathematical virtue and at the same time the central philosophical challenge. It falls to the interpretations to give content to this semantic indeterminacy by connecting the formalism with phenomena, models and epistemic states.

Initial probabilities

This point can be stated with practical precision. Applying probability calculus requires more than knowing rules of manipulation. One needs initial probabilities, priors or prior probabilities in the technical Bayesian sense: probabilities that precede the specific evidence at issue, though not necessarily a priori in the stronger philosophical sense. What is required is an assignment of probabilities to basic events, hypotheses or possible outcomes before any subsequent calculation.

With an ideal die, each face is said to have probability 1/6. With an ideal pack of cards, each card is said to have probability 1/52. But why is the object treated as ideal? That idealisation already involves a judgement of symmetry and the principle of indifference, which is itself a particular interpretation.

In real problems, one may say that a patient has a 20% risk of developing a condition, that a model has posterior probability 0.8 of being adequate according to a chosen criterion, or that a rare event has a chance of 1 in 10,000. But how are these numbers obtained?

This is where interpretation enters, because there is no single universal answer.

In some cases, probabilities are obtained from symmetry, as in the classical approach. In others, they are obtained from observed frequencies within a reference class, as in frequentism. Some contexts appeal to a causal or physical model intended to supply chances, as in dispositional variants. Other approaches derive them from normative criteria of rational belief, including coherence and principles for choosing initial probabilities, as in subjectivist and Bayesian traditions. In particular approaches they may also be obtained from informational constraints, such as maximum entropy, when the aim is to represent ignorance in a disciplined manner, as in strands of objective Bayesianism.

There are also cases in which these resources are combined. In every case, a conceptual decision is made about what counts as a legitimate basis for assigning probability.

The problem of initial probabilities is ultimately a problem of justification. What entitles us to assign this number rather than another? And why should that entitlement be recognised as rational, scientific or objective?

In some applied controversies this becomes explicit. In clinical trials, for example, the choice of initial probabilities may be disputed. Should they be based on historical data, deterministic models, simulations or expert opinion? Each answer carries a different conception of what confers legitimacy on a degree of belief or an estimate of risk.

The interpretative dispute is therefore not peripheral. It governs the first step, and the first step governs the rest. In the end, the question is how a number is connected to reality and to knowledge.

Interpretative map and comparative protocol

The book follows a simple strategy. First, it reconstructs the emergence and historical expansion of the concept, because interpretations do not arise in a vacuum. They respond to concrete problems, concrete uses and concrete crises, from the gaming table to the laboratory, from demography to physics.

Next, the modern formalism and what it stabilises are presented, with emphasis on the fact that stabilising the calculus is not the same as fixing its meaning. The book then moves into the comparative core.

The interpretations discussed can be organised into families, although the boundaries between them are porous.

The classical interpretation, associated with Pierre-Simon Laplace (1749–1827), connects probability with equiprobable events and symmetry, relying on principles such as indifference.

Frequentism, developed canonically by Richard von Mises (1883–1953) and reformulated pragmatically by Hans Reichenbach (1891–1953), defines probabilities as limiting relative frequencies in long sequences and therefore privileges repetition and the reference class.

The propensity interpretation, proposed by Karl Popper (1902–1994), seeks to preserve objectivity without reducing everything to frequency, treating probabilities as causal dispositions of experimental set-ups, with particular attention to singular events and quantum mechanics.

Best-system interpretations, associated above all with David Lewis, offer another analysis of objective chance. Instead of treating chance as a local power of an experimental set-up, they seek to identify it with the probabilities appearing in the probabilistic laws of the system that best organises the history of the world according to criteria such as simplicity, explanatory strength and fit. This family shares with propensity theory an interest in objective chances, but differs from it in its ontology and in the way those chances are grounded (Lewis, 1994; Hoefer, 2019).

The logical interpretation, defended by Rudolf Carnap (1891–1970), understands probability as a logical relation of confirmation between hypotheses and evidence, sustaining the aspirational idea of a logically correct value given a body of information and a fixed formal language.

Subjectivism, developed by Frank Ramsey (1903–1930) and Bruno de Finetti (1906–1985), understands probability as the degree of belief of a rational agent, requiring coherence and providing a theory of updating, frequently Bayesian, for learning from data.

Each interpretation will be assessed by its ability to deal with persistent problems such as the single case, dependence on infinite sequences and idealisations, the relation to practical decision, treatment of the reference-class problem and, above all, the way in which initial probabilities are justified.

The first part reconstructs the historical journey from probability as the art of conjecture to probability as a scientific language, while also presenting modern formalism and its semantic underdetermination.

The second part examines each interpretation in the context of its emergence and development, with symmetrical treatment and explicit criteria.

The third part tests these interpretations in cross-cutting debates and applied domains, at the boundaries where the use of probability reveals its hidden commitments.

This introduction therefore has a dual function. It delimits the problem and provides the criteria that will be used throughout the book.

Probability will be treated as a concept spanning calculus, world and epistemic state, whose formal unity coexists with interpretative plurality.

The book begins here, at this threshold between technique and metaphysics, because it is here, before any calculation is made, that one decides what one really thinks is being measured when a probability is assigned: an aspect of the world or a stage in knowledge.

1 Duality of Probability

(From the calculus of wagers to scientific theories, seventeenth to nineteenth centuries, with an epilogue in the early twentieth century)

The idea of probability emerged at gaming tables and, two centuries later, reached the heart of scientific explanation. That trajectory helps to explain why its meaning remains contested. When chance appears in public, it rarely presents itself as an abstract riddle. It appears as dispute, loss, gain, decision and risk.

An intuitive description is almost unavoidable. In a didactic register, probability appears as a quantitative measure between 0 and 1, with values close to 0 indicating highly improbable events and values close to 1 indicating highly probable events. In ordinary language, 0 is said to correspond to the impossible and 1 to the certain. This intuition is useful as a starting point, but it already contains a conceptual choice. The point is not to abandon it, but to make it open to scrutiny.[1]

The opening question is not 'what is probability?' but something more direct. How should one divide fairly, wager without deceiving oneself, or decide when information is incomplete? Philosophy enters when we realise that the same number may have different ontological and normative statuses. In one context, it seems to measure an objective chance associated with an event or a physical arrangement. In another, it seems to register an agent's epistemic state in the light of evidence. The concept begins as a response to a practical problem and grows into a theory of rationality under uncertainty.

This chapter reconstructs that path. It begins with games of chance, consolidates the idea of expectation, introduces the notion of regularity in populations, reaches the syntheses of the eighteenth century, moves into the social domain in the nineteenth century, enters statistical physics and concludes with a necessary epilogue. Even though the stated period runs from the seventeenth to the nineteenth century, the quantum turn of the early twentieth century must appear at least as a shadow. It makes explicit what had previously been capable of appearing as mere method or convenience.

[1] On certainty: the intuition '0 = impossible' and '1 = certain' is useful in discrete models and as a didactic starting point. In continuous spaces, however, there are possible events with probability 0, and probability 1 expresses certainty in the sense of the model ('almost surely'), not an absolute logical guarantee.

By the end, it becomes clear why the interpretative dispute is not a late academic nicety. It appears with the concept itself, because the concept arose to serve different tasks.

1.1 Games and quantification

Antoine Gombaud, the Chevalier de Méré (1607–1684), was not a professional mathematician. He was an attentive gambler, familiar with the psychology of risk and the seduction of luck. His role is decisive precisely for that reason. He represents the setting in which chance became a question of fairness and rationality without requiring metaphysics. When he asks how many attempts make it 'reasonable' to expect a certain result, as in the problem of the double six, chance ceases to be fate and becomes an accounting of possibilities.

The historical scene contains an instructive detail. De Méré moved in the courtly world of Louis XIV (1638–1715), and this matters philosophically. Probability emerges in connection with public practices of dispute and legitimacy. The question is not only 'what is the chance?', but also 'what is the fair division?', 'what is the rational wager?' and 'what criterion can be defended before others?'.

Two features remain at the root of the concept. One is repetition. Even if a wager is singular in the gambler's life, the structure of the game is repeatable. The other is the idea of fairness. 'Fair' is not merely 'probable'. In terms of mathematical expectation, fair is that which gives no one a systematic advantage. Here, chance is domesticated by a public and quantifiable rule, a criterion that can be defended before others. The germ of the calculus and the germ of the norm arise together.

1.2 World and norm

Blaise Pascal (1623–1662) and Pierre de Fermat (1601–1665) enter the story in response to a need that is both practical and theoretical. In their 1654 correspondence on problems of games of chance, the passage from guesswork to calculation crystallises. The issue is not merely to solve a game. It is to justify why the solution is rational, especially in the problem of a fair division of stakes when a game is interrupted before it is completed.

The historical importance of Pascal and Fermat does not lie in their having been the first to engage with questions of chance. There were earlier works and curiosities, and problems connected with games had already been addressed by other authors. The decisive step was the systematic approach, which extended the reach of probabilistic reasoning beyond games and made it a general model for thinking about non-deterministic processes.

The combinatorial solution can be read in two ways. It can be understood as revealing a mathematical structure inherent in the game, something that would be there independently of the agent. Or it can be understood as a rational protocol for deciding under ignorance, a norm for division when the future is unknown. The duality appears early and is not resolved by history. It merely becomes sharper.

This point gains depth when one notices that, in Pascal, probability does not remain confined to games. In Pensées, published posthumously in 1670, probabilistic language enters existential decision in the form of the celebrated wager, that is, in a domain without a clear frequentist basis. This does not legitimise every use of probability, but it helps to explain why probability escapes pure technique. Once it enters as a premise in consequential decisions, discussing its meaning ceases to be a caprice and becomes a theoretical necessity. Classical objections to the wager, such as the appeal to fear, the cost, the plurality of gods and insincere belief, make the philosophical point visible. The calculation may be impeccable, while the link between number, belief and decision remains debatable.

1.3 Decision and expected value

Christiaan Huygens (1629–1695) consolidated mathematical expectation, that is, expected value, as a criterion for evaluating wagers. Probability acquires an explicitly normative role here. It is not merely a calculation of combinations. It is a rule for decision. The link between probability and utility appears as part of the practice itself from the outset.

This development brings both power and cost. Its power lies in offering an apparently universal criterion. Given possible outcomes, probabilities and gains, one chooses what maximises expectation. Its conceptual cost lies in assuming a simple relation between value and money, an assumption that would later require refinement through the idea of diminishing marginal utility. The philosophical gain from this historical episode is clear. Probability is not merely a descriptive number. It operates as a norm of choice, a technology of rationality under risk.

1.4 Regularity and the law of large numbers

Jakob Bernoulli (1654–1705) shifts the horizon of the subject. Probability ceases to be merely a tool for wagers and becomes a language for understanding regularities in populations. In Ars Conjectandi, published posthumously in 1713, the weak law of large numbers appears as the first major limit theorem. The expression 'moral certainty' names a kind of certainty that is not absolute but is sufficiently robust to guide practice, science, law and deliberation. Chance remains in individual events, but stability emerges when the price of large numbers is accepted.

The law of large numbers is not merely a technical result. It proposes an epistemological scheme. Frequencies observed in series under relevant and sufficiently similar conditions tend to stabilise around a value as repetition increases. This creates a bridge between the uncertain and the reliable. And that bridge opens the interpretative dispute. Stability may be read as a property of the phenomenon, with the limiting frequency taken as an operational criterion for objective chance. It may also be read as a property of the method, that is, as a guarantee about the behaviour of a procedure.

Within the Bernoulli family, the subject broadens. Nikolaus Bernoulli (1687–1759) and Daniel Bernoulli (1700–1782) developed results connected with mathematical expectation and began the study of distributions of observational errors. This line reached a high point with Carl Friedrich Gauss (1777–1855) in the early nineteenth century, when measurement error and the typical form of fluctuations became central to a science that measures in order to know.

A decisive step enters here. Abraham de Moivre (1667–1754) shows that it is not enough to say that stability exists in the limit. One must understand the form of fluctuations before the limit, where life and science actually operate. By studying approximations to counting distributions, he inaugurates a geometry of chance. A population does not merely stabilise. Under certain conditions, it tends to assume typical forms.

The philosophical value of this episode lies both in the confidence it warrants and in the conditions it requires. If fluctuations have an approximately known form, uncertainty can be quantified, margins anticipated, deviations compared and error criteria constructed. At the same time, it becomes clear that probability advances through idealisations. What is gained in generality is paid for in assumptions. Confidence in the typical form of fluctuations depends on accepting that the real phenomenon conforms, with sufficient approximation, to an ideal model. This step is not empirically innocent. It is a point of contact between calculus and world and therefore an interpretative point.

1.5 Determinism and measurable ignorance

Pierre-Simon Laplace (1749–1827) is both the great systematiser of the classical interpretation and the architect of its central paradox. He begins from a universe governed by deterministic causality and elevates probability, the child of human ignorance, to the status of a central scientific tool. The tension between ontological determinism and epistemic probabilism profoundly marks the classical tradition.

The Laplacian instrument for measuring ignorance is the principle of indifference. In the absence of reasons for distinguishing cases, equal weights are assigned. The strength of the principle is evident. It makes it possible to construct initial probabilities when the world does not supply ready-made frequencies. Its weakness is equally evident. To apply it, one chooses a space of possibilities and a parametrisation, and that choice is rarely neutral. Change the representation and the 'indifferent' weights change with it. This is why the controversy surrounding the classical interpretation is not a matter of taste. It arises from the representational character of the starting point.

This is the historical moment when the expectation that 'probability' ought to have a single dominant meaning gains strength. A symbolic landmark of that ambition appears in Laplace's Mémoire sur la probabilité des causes par les événements (1774), where the language of the 'probability of causes' seeks both to discipline inference and to reinforce the idea of semantic unification. The problem is that the history of the concept itself pushes in different directions. A project of uniqueness may be elegant, but it must contend with the plurality of tasks that probability had already assumed.

1.6 Inference and initial probabilities

Thomas Bayes (1701–1761), through the mediation of Richard Price (1723–1791), shifts the centre of gravity of the subject. Probability ceases to serve only the prediction of effects and begins to support inference about causes, that is, about hypotheses. The idea of rational updating emerges clearly. In the light of evidence, one's degree of confidence in a hypothesis is adjusted. Bayes's rule condenses this move into an expression that would become foundational.

This advance increases rather than reduces the importance of the problem of initial probabilities. Updating requires a starting point. And to start is to assign probabilities before the specific evidence at issue, that is, initial probabilities in the technical sense. Interpretative conflict becomes unavoidable. Are these initial probabilities symmetries, frequencies, beliefs, objective chances, useful conventions or some compound of all these? Philosophically, the problem is one of justification. What entitles us to assign this number rather than another? The first step governs the rest.

Here, with particular clarity, lies the point at which the duality becomes unavoidable. If probabilities are assigned to hypotheses, the question of the status of the number changes register. Is it intended as a measure of something in the world, such as objective chance, or as an ordering of access to the world, a rational norm of belief and decision given the evidence? The interpretative conflict is not peripheral. It governs the very possibility of saying that a hypothesis has become more or less plausible because of data.

1.7 Social statistics and legitimation

Nicolas de Condorcet (1743–1794) opened the way for ‘social mathematics’ and for the application of probability to morals and politics. Adolphe Quetelet (1796–1874) extended this programme by organising population regularities and treating statistical distributions as tools for describing social phenomena. Probability now migrates into collective life: votes, courts, demography, insurance and public policy. Society comes to be seen as a domain in which regularities emerge from multitudes and in which public decisions seek probabilistic legitimation.

The promise is powerful. Populations appear to offer stability. The risk, however, changes in kind. The 'objectivity' of social statistics does not coincide with the objectivity typical of physical theories. It depends on conventions of classification, recording criteria, operational definitions and institutional choices. What is counted, how it is counted, who enters a class and who remains outside it: all of this affects the number. The reference-class problem appears here as a structural challenge rather than a technical detail.

Over the course of the nineteenth century, this strand develops into modern statistics and techniques of correlation and regression associated with the work of Francis Galton (1822–1911), Karl Pearson (1857–1936) and Raphael Weldon (1860–1906). At this point, probability begins to support methods for assessing hypotheses in the light of experimental data, and the theme of the 'strength of evidence' returns with new force. The philosophical question is reconfigured. Does the probabilistic number describe a regularity in the world, the reliability of a procedure, or a norm for accepting and rejecting hypotheses?

At the beginning of the twentieth century, this line becomes denser with Ronald Fisher (1890–1962), Jerzy Neyman (1894–1981) and Egon Pearson (1895–1980). Their statistical programmes are not identical: Fisher emphasises significance, estimation and information in the data, whereas Neyman and Pearson formulate an explicit theory of tests and decisions with long-run error control. In both cases, probability comes to support public procedures of inference, but the role assigned to the probabilistic number differs across the traditions.

1.8 Statistical physics and collective phenomena

The nineteenth century takes probability to a new level. It ceases to be merely an instrument for mastering measurement error and becomes a constituent of physical theories. Robert Brown (1773–1858) makes a phenomenon of microscopic fluctuation visible. The kinetic theory of gases and statistical thermodynamics then develop through the work of James Clerk Maxwell (1831–1879), Ludwig Boltzmann (1844–1906) and Josiah Willard Gibbs (1839–1903), with formalisations in terms of statistical ensembles.[2]

[2]A statistical ensemble in statistical mechanics is an idealised collection of copies of a system, under the same macroscopic constraints, used to define averages and distributions.

This domain shows particularly clearly the coexistence of distinct conceptions of probability. At some points, probability appears epistemic, a way of dealing with unknown microstates. At others, it appears pragmatic, a shortcut in the face of intractable complexity. At still others, it appears almost ontological, because probabilistic vocabulary becomes indispensable to macroscopic description: the appropriate description of collective phenomena in which macroscopic quantities acquire their meaning precisely as statistical descriptions.

The difference is not merely semantic. It affects what counts as explanation. The same theory permits different readings of the role of chance. This prepares the ground for the twentieth century. It shows that probability can be more than ignorance. It can be the appropriate language for certain levels of description. The duality reappears here as a dispute about what counts as explanation.

1.9 Quantum foreshadowing

Although the quantum turn belongs to the early twentieth century, it is worth introducing it as an epilogue because it makes unavoidable what had previously been capable of appearing as methodological convenience. Between 1905 and 1906, Marian von Smoluchowski (1872–1917) and Albert Einstein (1879–1955) consolidated the analysis of Brownian motion, giving chance a connection between microstructure and observable phenomenon.

The transition to quantum physics occurred in stages that should be distinguished. In 1900, Max Planck (1858–1947), in addressing black-body radiation, introduced the quantisation of energy as part of a solution to a problem in radiation theory. In the years that followed, Albert Einstein extended the break with classical physics in work on light quanta. Quantum mechanics proper was formulated chiefly in the 1920s by figures such as Werner Heisenberg (1901–1976), Erwin Schrödinger (1887–1961), Max Born (1882–1970), Paul Dirac (1902–1984), Louis de Broglie (1892–1987) and Niels Bohr (1885–1962). It is at this stage that probability acquires a structural position in the prediction of measurement outcomes, without this fact by itself yielding a unique philosophical interpretation of the theory.

The philosophical effect is immediate. If the formalism supplies probabilities for measurement outcomes, the question ‘probability of what?’ cannot be postponed. Are these objective chances? Limitations of knowledge? Rules for rational expectations in the face of an experimental apparatus? The point here is to show that the historical genesis converges on a point at which meaning can no longer be hidden behind technique.

1.10 Synthesis

The history reconstructed in this chapter is not a linear march towards clarity. It is the chronicle of the forging of an intrinsically divided concept. Chance enters mathematics through a practical route, games and divisions of stakes, and soon demands a public norm of decision. The ideas of population and regularity then make the number appear as a portrait of objective stability. Inference about causes subsequently makes the problem of initial probabilities unavoidable. Social administration reveals how statistical 'objectivity' can be shaped by institutional criteria. Statistical physics shows that probability can operate as an appropriate language of collective phenomena. Quantum theory, finally, prevents the question of meaning from being treated as a merely methodological detail.

No single interpretation can embrace all these historical tasks without paying a high conceptual price. The classical interpretation captures symmetry but stumbles over dependence on representation. Frequentism inherits the promise of stabilisation but faces idealisations and the reference-class problem. The inferential line opens the domain of hypotheses but exposes the justification of initial probabilities as a point of cost. Social statistics expands domain and power but makes the role of conventions visible. Physics, from collective phenomena to quantum theory, presses the concept towards a reading in terms of objective chance while keeping the epistemic alternative alive.

For this reason, the historical genesis offers no definitive answer to the question 'what is probability?'. It exposes the genealogy of the conflict. If, in the twentieth century, a unification of the calculus is consolidated through Andrei Kolmogorov's (1903–1987) axiomatisation, it will be a mathematical victory rather than a philosophical resolution. The formalism organises, but does not choose. It falls to the interpretations to give content to the number and to state what is being assumed when a probability is assigned.

With that diagnosis, the book moves on. The next chapter turns to modern formalism and the interpretative neutrality of the calculus, before returning, with explicit criteria, to the map of competing interpretations.

2 Formalism and neutrality

(How axiomatisation stabilises the calculus without fixing objective chance or degree of belief)

If the previous chapter showed how probability developed through heterogeneous problems, this chapter examines how that multiplicity required and received a formal unification which, paradoxically, made interpretative diversity more conspicuous. Before receiving a unified formal syntax, the semantics of probability had been forged by heterogeneous problems. In different traditions, probability operated as a criterion for division and wagering, matured into a promise of regularity in populations, expanded into an instrument for inferring causes and, in the nineteenth century, entered physics and social statistics as a language of collective phenomena. The calculus grew faster than agreement about what it meant. The demand for foundations therefore became unavoidable. What was sought was a structure capable of accommodating coins and continuous functions, urns and temporal processes, frequencies and inferences, without depending on fragile intuitions.

Andrei Kolmogorov's (1903–1987) axiomatisation of 1933 answers that demand. It offers an abstract axiomatic structure. A mathematical object is defined, conditions are imposed so that P is a normalised, σ-additive measure and coherence of the calculus follows. The move is powerful and, at the same time, philosophically provocative. It solves the problem of how to calculate probabilities in a unified way, but it does not fix the bridge between formalism and application. It does not decide what P represents in a given context, whether objective chance, degree of belief or an explicitly stated hybrid position, and it does not determine how Ω, $\mathcal{F}$ and P are to be selected for a concrete system. The calculus stabilises the procedure. Meaning remains an interpretative task. This separation is not a detail. It is the key to the twentieth century. The neutrality here is structural. The formalism underdetermines the semantics, and for that reason rival interpretations can inhabit the same syntax without collapsing into one another. Nor is degree of belief a mere guess. It is an epistemic state subject to norms of coherence and may be rationally constrained by symmetries, data and losses.

2.1 Crises and paradoxes

The classical theory associated with symmetry has always carried an elegant promise. If there is no reason to distinguish between possibilities, the possibilities are to be treated as equally probable. The problem is that, outside highly controlled examples, it is rarely clear which parametrisation induces the relevant notion of uniformity. The same situation can be described in different ways, and the principle of indifference can yield different answers to the same question expressed in ordinary language.

Joseph Bertrand's (1822–1900) paradoxes made this weakness explicit. In a celebrated example, one asks for the probability that a chord chosen ‘at random’ in a circle is longer than the side of an inscribed equilateral triangle. At first sight, the question appears well defined. Yet different, equally plausible methods of choosing ‘a chord at random’ lead to different probabilities. The paradox is not merely a puzzle. It reveals a structural fact. The expression ‘at random’ is not a mathematical primitive, in the sense that the formalism does not define it without a generating rule. It is a vernacular invocation which, in order to be operationalised, requires a model of randomness: a generating rule, a choice of a space of possibilities and therefore an implicit commitment as to what is to count as uniform. The paradox is less about chords than about the fact that ‘at random’ is a hypothesis about a procedure.

This kind of crisis is philosophically decisive because it dismantles a naïve expectation. It is not enough to invoke objectivity. The criterion at issue must be made explicit: for example, invariance of the procedure under relevant transformations as a justification of objectivity in the sense of independence from the agent, or the anchoring of the assignment in an objective chance of the physical set-up. In the classical case, objectivity is often said to arise from symmetry. Bertrand shows that symmetry depends on how the sample space is represented. Change the representation and the relevant symmetry changes. Assigning probabilities requires both a principle and a representation that induces a measure. Without that choice, there is no determinate notion of ‘uniform’.

At this point it becomes clear why twentieth-century axiomatisation was necessary, but also why it could not by itself end the debate. A rigorous formalism can prevent internal contradictions, but it cannot by itself decide which modelling adequately represents a concrete situation. Bertrand's paradox teaches a lesson that will run throughout this book. Interpretation enters before the calculation, at the moment when one decides what is being counted.

2.2 Axioms and probability space

Kolmogorov's (1903–1987) proposal states precisely what is required in order to carry out probabilistic calculation in general. It begins not with coins or frequencies but with an abstract structure. There is a set of possible outcomes in the model, a set of events about which one wishes to speak and a function assigning to each event a number between 0 and 1. That function must satisfy conditions that are both simple and powerful.

The first condition says that probabilities are non-negative. The second says that the certain event receives probability 1. The third says that, for any countable sequence of mutually exclusive events, the probability that 'one of them' occurs is the sum of their probabilities. This last condition, σ-additivity, is the heart of the calculus. It formalises the idea that probability behaves as a measure. Its countable form sustains the calculus in infinite settings and underpins arguments involving limits and approximations.

A probability space is usually represented by the triple $(\Omega, \mathcal{F}, P)$, where Ω is the sample space, understood here as the set of possible outcomes in the model. $\mathcal{F}$ is a collection of admissible events, a σ-algebra on Ω, ensuring that natural operations such as complements and countable unions remain within what counts as an event. P is the probability measure.

The requirement that $\mathcal{F}$ be a σ-algebra is not a mere formality. It embodies a methodological decision about the scope of what can be treated once the door is opened to infinitely many ideal possibilities. In continuous settings, not every subset can be treated as an event without technical impasses. By imposing closure under countable operations, the formalism ensures that one can reason systematically about infinite families of events. A legitimate 'event' is assumed, in the end, to be one that can be integrated into the economy of the calculus without consistency breaking down.

The gain is evident. A single formalism covers discrete and continuous cases, simple experiments and complex processes. It allows random variables to be defined as functions and distributions to be treated as induced measures. It organises conditioning in a general way. It creates the setting in which limit theorems can be stated precisely. From this point onwards, probability is no longer a collection of methods. It becomes a unified theory.

The philosophical cost comes with the gain. The formalism does not fix the correspondence between an empirical system and the sample space Ω. Nor does it determine which subsets should be treated as admissible events in $\mathcal{F}$, or provide a unique criterion for selecting the measure P in an application. It says only how P must behave if the calculus is to be coherent. Axiomatisation solves the problem of internal coherence. It does not choose an interpretation. That neutrality is part of its power.

2.3 Conditional probability, independence and variables

With a probability space in hand, operational concepts take precise form. They are operational because they appear constantly in scientific practice. They are the means by which probability enters inference, prediction and explanation.

Conditional probability answers questions of the form ‘given that I know this, how probable is that?’. In elementary formalism, for events A and B with P(B) > 0, one defines:

$$P(A \mid B) = \frac{P(A \cap B)}{P(B)}.$$

The expression is simple, but it carries a requirement. An initial measure P is needed for the conditional probability to make sense. One updates from what one already has. Conditioning by itself therefore does not solve the problem of initial probabilities. It makes that problem more visible. In continuous settings, the idea of conditioning requires additional care, because one does not always condition on events with positive probability.

The definition is mathematically uncontroversial, but the reading of ‘given that’ is not. On an epistemic reading, one conditions in order to update beliefs in the light of evidence, and it is here that the Bayesian tradition finds its normative core. On a more operational and sometimes frequentist reading, conditioning is a way of restricting the analysis to a subpopulation, a stratum or a set of experiments in which condition B is satisfied. The formula is the same. What changes is the ontology of what is being manipulated: beliefs or properties of a procedure.

Independence is another central concept. Events A and B are said to be independent when $P(A \cap B) = P(A)P(B)$. Operationally, independence makes it possible to decompose complex problems into parts, thereby making calculations and models feasible. Philosophically, independence is rarely a brute empirical fact. It is often a methodological idealisation. Sometimes it is a hypothesis about the world. At other times, it is a modelling assumption adopted to make the world tractable. Independence may be assumed because the dependence structure is unknown, because the dependence is small enough to be neglected, because one is working with approximations, or because the model becomes intractable without that postulate. To assert independence is to assert something about the world or about a modelling strategy, and often both are asserted without being distinguished. This tension will run, in different forms, through all the twentieth-century interpretations.

Random variables provide the bridge between events and measurable quantities. A random variable is a function that associates a number with each possible outcome. The definition appears straightforward, but it is the formal way of saying that an observable quantity is extracted from chance. The formalism makes it possible to treat distributions, moments, transformations, sums and limits. Above all, it makes it possible to treat noise and error in a disciplined way, as part of a language of uncertainty that can be calculated.

A simple example shows how the formalism enters practice. In a diagnostic test, a patient may or may not have a condition, and the test may be positive or negative. The relevant question is almost never 'what is the probability that the test is positive?'. It is more often 'what is the probability that the condition is present given a positive test?'. The difference is between a rate belonging to a procedure and an inference about a hidden state. Conditional probability becomes unavoidable. Even so, the final number depends on base probabilities such as prevalence and the characteristics of the test. The calculation proceeds firmly. The justification of the starting point remains an interpretative and methodological question.

2.4 Limit laws and implications

Paradoxes such as Bertrand's revealed weaknesses of representation, while axiomatisation offered formal stability. But could large-scale behaviour provide the definitive bridge between formalism and world? The answer depends on how one interprets the role of limits and idealisations themselves. Two results organise modern confidence in the scientific use of probability: the law of large numbers and the central limit theorem.

Conceptually, the law of large numbers says that relative frequencies approach probabilities as the number of repetitions increases, under appropriate conditions. It gives content to the idea that probabilities are something manifested in the long run. It is often treated as a bridge between world and calculus. But the bridge has two faces. On an ontological reading, it suggests that there is a real chance which reveals itself as regularity. On an epistemic reading, it says that an estimation method tends to work well in the long run. It is one thing to say that the world has a rate. It is another to say that, under the modelled conditions, a procedure converges to a value. The theorem itself does not choose between these readings.

There is a further nuance that is not merely technical. The weak form of the law guarantees convergence in probability. The strong form guarantees almost-sure convergence. The difference may appear subtle, but it touches a philosophical nerve. Almost-sure convergence does not mean 'absolute certainty' in ordinary language; it says that the set of trajectories on which convergence fails has probability zero. By introducing this vocabulary, the theory makes explicit the kind of idealisation it accepts. It speaks of infinite sequences, asymptotic behaviour and exceptions of measure zero. Probability becomes a language of 'almost always', and this 'almost' is mathematically disciplined but is not the same as the 'never' of ordinary language.

The central limit theorem explains why the normal distribution appears so frequently in errors and in sums of many contributions. Conceptually, it says that, in many situations, the sum of many small and reasonably well-behaved contributions tends, when appropriately rescaled, to produce an approximately normal distribution. Philosophically, this supports an idea of robustness. Even when the exact distribution of each contribution is unknown, it is often possible to work with a typical approximation.

That robustness, however, has conditions. It can fail under strong dependence, heavy tails, when a few factors dominate the population, or when the context violates the assumptions of the theorem. The central limit theorem thus becomes part of the debate about idealisation. It shows why models can work even when they are not literally 'true', but also why trusting them requires attention to the conditions of applicability.

The broader implication is that limits and operations of composition introduce a kind of objectivity through stability. Something becomes tractable not because each individual event is predictable, but because large collections display patterns. Here probability approaches a grammar of regularity. And this move, which appears purely mathematical, is also epistemological. It informs both what one should expect from the world and what one may demand of a method.

2.5 Calculation and meaning

With axiomatisation, probability acquires a grammar that works almost everywhere. That universality can create an illusion: because the calculus is unique, one may imagine that its meaning is unique as well. The central philosophical point of this chapter is to reject that illusion precisely.

The formalism is compatible with different readings because it is deliberately a theory of structure, not a theory of empirical meaning. It states how P must behave in order to remain non-contradictory when events are combined, when conditioning takes place and when quantities are aggregated. It does not say whether P represents a limiting frequency, a coherent degree of belief, a logical relation of confirmation, an objective chance associated with a physical set-up or some hybrid. It permits all these readings provided that they respect the rules.

This deliberate abstraction is partly analogous to the separation between formal geometry and its physical application. Kolmogorov's axioms provide the dominant framework for the real-valued, point-valued and σ-additive probability treated in this book, but they do not exhaust every possible formalisation of uncertainty. Within this standard framework, distinct philosophical interpretations may share the same calculus without the axioms determining which semantics must be adopted. The law of large numbers, for example, may be read as objective regularity, as a property of learning procedures, as a result compatible with subjectivist models based on exchangeability, or as a manifestation of physical chances. The formula remains; the interpretative commitment varies.

We can therefore return to Bertrand's paradox with greater clarity. The formalism does not go wrong. It simply does not say which measure should represent the 'random choice' of a chord. The answer depends on the generating mechanism and on how possibilities are represented in the model. Interpretation does not enter at the end as a commentary. It enters at the beginning as a condition of application.

The same is true in inference. When Bayes's rule is applied, one conditions. But conditionalisation begins only after a base measure has been chosen. That starting point may be defended by symmetry, frequencies, informational constraints, subjective coherence, causal models or a combination of these resources. The calculus does not choose among these defences. It requires only consistency once the choice has been made.

It is sometimes said that axiomatisation made probability 'purely mathematical'. This is correct in one sense and misleading in another. It is correct because the formalism can be studied without commitment to a specific interpretation. It is misleading because applying the formalism to the world always requires an act of connection. One chooses the sample space, what is to count as an event and a measure; one decides what conditioning and independence mean in the context. That act of connection is precisely where philosophy enters.

2.6 Synthesis

Kolmogorov's (1903–1987) formalism is a mathematical achievement of great reach. It unifies the calculus, integrates probability with measure theory, provides a rigorous foundation for limit theorems and stabilises general conditions of coherence. It solves the problem of how to calculate safely and enables different fields to share the same conceptual infrastructure.

Yet this formal achievement illuminates rather than dissolves the interpretative problem. Axiomatisation does not decide how to model 'random choice' in Bertrand's paradox. It does not arbitrate between ontological and epistemic readings of limit theorems. It does not justify choices of base measures in inference. Nor does it determine the philosophical status of independence, which so often functions as a methodological idealisation. It offers a powerful syntax, but delegates semantics to philosophy.

In sum, axiomatisation provides the formal rules of the calculus, but it does not draw the map. It falls to the interpretations to justify the first line drawn, the zero point from which the entire probabilistic construction rises. From this point onwards, interpretations appear as responses to the semantic space left open by the axioms, a space that is not a flaw in the formalism but the condition of its generality. Each tradition will seek, in its own way, to justify the passage from abstract formalism to concrete probability assignments. That is what the following chapters will examine: not as a mere conceptual dispute, but as an investigation of the point at which a number acquires warrant, that is, the point at which one decides what one thinks one is counting when assigning a probability.

3 Classical Interpretation

(Indifference, symmetry and justificatory circularity)

The classical interpretation marks a historical beginning and, at the same time, makes a conceptual limit explicit. It introduces an initial rule intended to provide objectivity without appealing to long series of observations. Yet that very rule exposes a structural limit when one seeks to derive, solely from the absence of information, an assignment of probability through P that is intended to be unique and non-arbitrary, without making explicit whether that assignment represents an objective chance in the world, a degree of belief in the agent or, rather, an artefact of a modelling decision about how ignorance is to be represented in the probability space $(\Omega, \mathcal{F}, P)$. By 'reference measure in the model' we mean the base measure μ that stabilises equipossibility and, in continuous settings, underlies densities, such as Lebesgue measure on $\mathbb{R}^n$ or counting measure on discrete spaces. This chapter does not yet fix the final semantics of P. It examines the methodological claim that unique initial probabilities can be obtained from indifference alone, without declaring the commitments that fix the space, individuation and measure.

The central thesis of this chapter can be stated with conceptual precision. The classical interpretation incurs justificatory circularity in grounding initial probabilities. It promises that objectivity, understood as independence from individual epistemic states and obtained through an intersubjectively accessible and applicable rule of initial assignment, emerges from the absence of reasons to prefer one alternative. Yet, in order to apply its central principle, it must decide beforehand what that principle was supposed to justify. It must choose a space of possibilities, fix what counts as a distinct possibility and determine under which measure two possibilities are to be treated as 'equal'. Without those decisions, there is nowhere for indifference even to be applied.

The circle is more than logical. It is not merely a universal formal vice. It is constitutive whenever uniqueness is promised 'from ignorance', as though the absence of reasons were sufficient, when in fact the application of the principle depends on a representational structure already fixed. It reveals that classical 'objectivity' is not a discovered foundation but a stabilised product of the interaction between a rule, indifference, and a prior representational structure, space and measure. Classical objectivity arises not from ignorance taken in isolation, but from a combination of the absence of relevant information and a prior rule that partitions the space into alternatives and defines a reference measure in the model.

There is a structural analogy here with a broader problem in contemporary epistemology: the status of 'starting points'. Just as the legitimacy of basic beliefs, intuitions and initial criteria is debated in theories of knowledge, so here the legitimacy of basic spaces and reference measures in probability assignment is at issue. The form of the challenge is the same. If the first step already contains, implicitly, what was supposed to be concluded, the promised objectivity becomes only an appearance of objectivity, obtained by naturalising earlier choices.

This discussion is not intended to disqualify the classical interpretation. It is intended to delimit its strength, its legitimate domain and its legacy. The classical interpretation is indispensable because, already at the beginnings of systematic probabilistic theorising in the seventeenth and eighteenth centuries, it locates the problem of initial probabilities. In doing so, it identifies the point at which every interpretation enters, even when one prefers not to name it.

3.1 Indifference and representation

The heart of the classical interpretation is easy to state. In the absence of relevant reasons, given the available information and the explicitly defined scope of the model, the same weight is assigned to alternatives defined as equiprobable under the adopted measure. The difficulty lies in three expressions that appear innocent: 'there are no reasons', 'distinguish' and 'alternatives'.

'Alternatives' do not come ready-made. They must be constructed by delimiting the space of possibilities. To 'distinguish' requires a criterion of individuation, that is, a rule for deciding when two descriptions count as two possibilities and when they are merely two descriptions of the same case. 'There are no reasons' is a claim about relevant information, and relevance is rarely neutral. In finite and highly symmetrical cases, such as idealised games, this structure tends to remain invisible. In continuous domains or spaces of scientific hypotheses, it reappears as an unavoidable choice.

It is useful here to note a conceptual transition inherited by the classical interpretation even when it is not announced. In Gottfried Wilhelm Leibniz (1646–1716), the idea of not privileging without sufficient reason has a metaphysical and theological flavour. In the classical interpretation, the same form appears as a methodological rule for the human construction of models. The transition is decisive. Indifference ceases to be a thesis about reality and becomes a discipline of starting points, a way of organising what counts as 'equal' under a given delimitation. This is why the principle of indifference is not a theorem. It is a regulative principle at the methodological level. It prescribes an initial choice of measure and equipossibility when the problem, as modelled, does not distinguish alternatives by the available information. Its ambition is to produce public objectivity. Its limit is that it depends, silently, on representational decisions that already carry commitments.

3.2 Equiprobability as a postulate

In its best-known form, the classical definition connects probability with the ratio of favourable cases to possible cases, provided that the possible cases are equipossible. This formulation is usually associated with Pierre-Simon Laplace (1749–1827) and, through important combinatorial antecedents, with authors such as Jacob Bernoulli (1654–1705), in whose tradition the idea crystallises as a general rule. When nothing leads us to expect one case to occur more often than another, the cases are treated as equally possible. Within this framework, if elementary events receive equal probabilities, the probability of a disjunction of elementary events is simply the number of cases in the disjunction divided by the total number of possible cases.

Conceptually, the definition requires three conditions. First, the space must be capable of being treated as composed of comparable 'cases'. This is natural in discrete and artificial domains. It becomes problematic when the space is continuous or when the object is not an outcome but a hypothesis.

It also requires a criterion of equipossibility. In games, this is usually grounded in the symmetry of the device, but that symmetry is a hypothesis about the mechanism, supported by construction, inspection, testing and often auditing. Even in the 'pure' case of a die, equipossibility is not an a priori truth in the strong philosophical sense. It is a methodological idealisation that can be supported by empirical control of the device and by a model that treats it as an adequate approximation. Equipossibility, therefore, is not a metaphysical given but a model postulate pragmatically justified by empirical control and operational consensus. It is a condition for applying the principle, not a consequence justified by the principle.

It further requires the event to be well defined under the chosen delimitation. If the description changes the delimitation, the result changes. The classical interpretation works well when the description is stable and there is public agreement about the relevant delimitation. It loses force when the delimitation depends on contestable choices.

Here justificatory circularity becomes sharp. The classical interpretation wants indifference to produce the starting point. But the starting point exists only after the space and the reference measure in the model, within which indifference will be applied, have been chosen. It is no accident that, in the nineteenth century, authors such as John Venn (1834–1923) and George Boole (1815–1864) pressed this point, helping to shift the search for objectivity towards frequentist anchors. In the twentieth century, that shift found an influential methodological expression in Ronald A. Fisher (1890–1962), without implying any simple causal chain. Even so, the classical definition did not disappear. It returns by new routes when Bayesian probability becomes central and once again requires initial probabilities. In this reconfiguration, the principle of indifference reappears as a source of ignorance-based initial probabilities precisely because it promises a beginning without explicit preferences.

3.3 Bertrand's paradox and measure

Joseph Bertrand (1822–1900) made famous a kind of paradox that is no parlour game. It is a conceptual microscope. The lesson is simple and disturbing. In continuous problems, the expression 'choose at random' does not determine a unique procedure. Different procedures, equally plausible in ordinary language, can produce different probabilities.

The chord-in-a-circle case illustrates this. The question 'what is the probability that the chord is longer than a certain length?' receives different answers according to how 'at random' is operationalised. There are natural ways of choosing a chord by its midpoint, by its endpoints or by geometrical parametrisations. Each method specifies a distinct generating mechanism and, with it, a distinct notion of operational randomness. Each mechanism induces a different measure on the space of chords. The result is not noise. It is structure. It shows that, in continuous settings, equipossibility is not an intrinsic property of the abstract set of possibilities. It is a relation between possibilities and a reference measure in the model. The paradox reveals that 'equipossibility' is always relative to a specific generating mechanism. The reference measure in the model is the mathematical encoding of the assumption about how the random process operates.

When the classical interpretation aspires to generality, it needs a rule for choosing that measure. Bertrand shows that the ideal of a 'natural uniformity' fails unless the generating mechanism is specified. Static geometry is not enough. The expression 'at random' requires an operational specification of the procedure and of the reference measure it induces. One must say what is done, not merely what is imagined.

3.4 Technical a priori and reparametrisation

The rule of succession, associated with Pierre-Simon Laplace (1749–1827), is a didactic example of how the classical interpretation attempts to 'start from zero' in inference. Given observed successes, one wishes to estimate the plausibility of a further success. The classical procedure frequently assumes an initial 'uniform' distribution, as though uniformity were synonymous with neutrality.

The difficulty is conceptual. 'Uniform' depends on how the parameter is represented. The same absence of relevant information, represented as an initial measure on parameter space, can take different forms under reparametrisation, and 'uniformity' is not invariant. A distribution that appears neutral under one parametrisation ceases to appear neutral under another. This is not a technicality. It is evidence that there is no absolute neutrality without a privileged co-ordinate system. Ignorance is not a single content that can be poured, without remainder, into a single distribution. Where neutrality exists, it is a relation between distribution and description, and that relation depends on the representational system.

The connection with Bertrand is direct. In Bertrand's case, the problem appears as the choice of measure on a continuous space of possibilities. In the rule of succession, it appears as the choice of representation on a space of hypotheses. In both cases, the conclusion is the same. Indifference does not operate without a prior structure defining what counts as 'equal'. And that structure is not supplied by the principle itself.

3.5 Invariance and maximum entropy

Contemporary attempts to preserve the classical impulse tend to follow two strategies.

The first is the strategy of invariance. Instead of saying 'I have no preference', one makes explicit that, given certain information and a set of transformations regarded as irrelevant, the model should be invariant, and this constrains the choice of initial measure. The idea is to make neutrality depend on symmetries under transformations deemed irrelevant. This reduces arbitrariness, but opens an earlier question. Who decides which transformations are irrelevant? In some physical problems there may be a robust justification. In epistemic and inferential problems, the choice of an invariance group may be as contestable as the original choice.

The second strategy is maximum entropy, associated with Edwin T. Jaynes (1922–1998). Its impulse is to turn indifference into an informational rule. Given certain constraints imposed in the model, one chooses the distribution that maximises entropy subject to those constraints, interpreting this as a methodological rule for constructing an initial measure rather than as metaphysical neutrality. The proposal is powerful as a discipline, but it is not metaphysically neutral. It depends on the hypothesis space and, in continuous settings, also on what is taken as the reference measure in the model. In continuous settings, the rule is better understood as maximisation of entropy relative to a base measure, not as absolute neutrality. Maximum entropy therefore displaces the problem of choosing a measure, but does not eliminate it. The principle requires an underlying measure relative to which entropy is calculated. Choosing that measure, whether Lebesgue measure, counting measure or another, is once again an act of modelling not justified by the informational principle itself, and thus returns in a more sophisticated form to the problem of the reference measure.[1]

Thomas Seidenfeld (1947–) sharpened this point in his criticisms of the ideal of a 'purely objective' Bayesianism. There is no general and uncontroversial procedure that completely eliminates the role of initial judgements. They can be made explicit. They

[1] Lebesgue measure λ_n on $\mathbb{R}^n$ (Henri Lebesgue, 1875-1941) rigorously generalises length, area and volume. It is translation-invariant (and therefore compatible with the usual Euclidean symmetries, such as rotations) and satisfies $\lambda_n(\prod_{i=1}^n [a_i,b_i]) = \prod_{i=1}^n (b_i - a_i)$. In continuous models, it serves as a reference measure for densities: if $P \ll \lambda_n$, then $P(A)=\int_a f\, d\lambda_n$. Choosing λ_n as the reference is a modelling commitment, in the same sense in which Joseph Bertrand (1822–1900) makes dependence on measure explicit in continuous problems.

can be reduced. They can be better justified. But the hope of an absolutely impersonal beginning often reappears as the hope of a 'natural' choice at another level.

3.6 The legitimate domain of the classical interpretation

Games and lotteries appear to be the ideal territory of the classical interpretation. Yet it is instructive to notice why they work so well. They work because they are stabilised practices. Symmetry is constructed and preserved, institutional routines of control are in place and there is epistemic agreement on the relevant delimitation.

Stabilisation occurs at three levels. At the physical level, one seeks to ensure that the device behaves approximately symmetrically. At the institutional level, rules, audits and standards are defined to discourage biased mechanisms. At the epistemic level, a practical agreement develops about what counts as an alternative and what counts as an event. Objectivity, in the sense of independence from individual epistemic states, is procedural and institutional. It results from physical controls, public rules and agreement about delimitation, not from ignorance taken in isolation as a foundation. This objectivity is public and procedural. It does not by itself amount to an ontological guarantee of 'chance' independent of a model. Classical objectivity here reveals itself as hybrid in nature, combining epistemic elements in the agent's indifference, formal elements in the structure of the space and the reference measure, and social elements in stabilising practices that make the procedure replicable and open to scrutiny.

When one moves to complex and continuous events, the situation changes. Symmetry is rarely 'given'. It is inferred, idealised and modelled. The difference between games and complex domains is above all a difference in control. The less control there is, the more explicit the modelling must be and the more visible the role of initial choices becomes.

This corrects a common caricature. The classical interpretation is not irrational outside games. It simply ceases to be self-justifying. It must declare its commitments, show its stabilising mechanisms or accept that neutrality will be relative to a particular delimitation.

3.7 Synthesis

The classical interpretation does not fail because of a mathematical error. It fails as a universal and self-sufficient foundation because its central rule depends on what it is meant to ground. It needs a space of possibilities, a criterion of individuation and a reference measure in the model for 'equality' to have meaning. In finite symmetrical settings, these decisions can be publicly justified by construction and control. In continuous settings and spaces of hypotheses, they reappear as choices, and choices require criteria.

There is a methodological consequence here that applies throughout the book. To examine any interpretation of probability, it is useful to ask what foundational problem it promises to solve, which assumptions it makes explicit, which assumptions it naturalises and which new problems arise when its 'success' is taken seriously. The classical interpretation is paradigmatic because, once analysed, it exposes the machinery of the beginning and thereby turns a habit into an object of philosophical enquiry.

Its legacy is therefore greater than its claim. The classical interpretation correctly identifies the problem of initial probabilities and provides a regulative ideal that remains valuable: the ideal of beginning without illegitimate preferences. What it also reveals is that, in probability, objectivity is rarely a pure starting point. It tends to be the product of explicit practices, representations and constraints.

Frequentism will therefore emerge not as a denial but as a strategic shift. If the classical interpretation fails because its objectivity depends on prior representational choices, frequentism will attempt to anchor objectivity in empirical regularities. Yet that empirical anchor moves the problem to the choice of reference class, admissible repetition and conditions for the stability of the mechanism. The challenge, however, is symmetrical. Whereas the classical interpretation stumbles over the arbitrariness of the reference measure, frequentism will face the arbitrariness of the reference class. In both cases, the promised objectivity depends on choices that precede application of the method. The pattern is confirmed. Every interpretative solution turns one kind of problem into its point of strength at the cost of inheriting a new kind of conceptual cost. This is why the interpretative dispute is not historical noise. It is the very form of the problem.

4 Frequentist Interpretation

(When objectivity depends on the delimitation)

Frequentism did not arise as a school that simply 'prefers' one definition. It emerged as a response to a discomfort. The previous chapter showed why the classical interpretation promises too much when it tries to transform ignorance into objectivity through symmetry. The frequentist move is almost a methodological pledge. If probability is to be public, it must rest on observable regularities within an explicitly stated reference class, that is, under rules for identifying repeatable cases that can be audited and compared. It must rest not on a principle that distributes equipossibilities, but on a regularity in a population.

With this, the location of 'objectivity' changes. It ceases to be an attribute of a mind that apportions ignorance and becomes an attribute of a procedure that fixes delimitations, repeats conditions and makes its method of counting publicly visible.

This shift solves one problem and exposes another. Frequentism is strong where there are series, replications, protocols, populations and repeatable procedures. It struggles when the question arrives as a singularity: tomorrow's rain, the risk to this patient, the failure rate of a type of bridge under specified conditions applied to this particular crossing. The frequentist virtue lies in long-run methodological objectivity. Procedures are evaluated by their behaviour under repetitions conducted in specified conditions. The cost is that the single case requires a bridge through a reference class.

This chapter follows the trajectory of that virtue and that cost, from their early philosophical motivations to their statistical legacy as a technology of inference. One point should be fixed at the outset. Frequentism is first and foremost a semantic interpretation of P linked to long-run frequencies, not an automatic synonym for 'classical inference'. The publicity it promises is conditional. The delimitation, protocol, reference class and rule of repetition must be made public; without them, the number is not public but ornamental.

4.1 Why series matter

In Robert Leslie Ellis (1817–1859), the shift towards the population has a root that is not merely mathematical but philosophical. Laplacian probability appeared to him to depend on an implicit psychology. The classical interpretation assumes that, where there is no reason to discriminate, the mind should distribute equal weight. Ellis does not treat this as evidence of rationality. He treats it as a weakness. In Ellis's criticism, the absence of reasons, without an explicit delimitation and measure, does not determine a unique numerical assignment. Observable regularity, by contrast, is presented as an intersubjective datum rather than the product of an intellectual operation that distributes ignorance into weights. The point is not to deny the calculus. It is to distrust the path from 'I do not know' to 'it is one half' when the transition is made without declaring the space of cases and the rule of equalisation. In terms of intellectual responsibility, the point is severe. If equipossibility is not stabilised by practice, instrument or explicit convention, it is merely a disguised hypothesis. A series, by contrast, forces us to say how repetition was carried out, what was held constant, what varied and why this still counts as the 'same experiment'.

John Venn (1834–1923) makes the motivation still more explicit. In The Logic of Chance, he enters a concrete philosophical debate. Historical determinism, associated with Henry Thomas Buckle (1821–1862), treated stable social statistics as evidence of rigid necessity in human conduct. Venn accepts that the physical world may be deterministic. What he rejects is the metaphysical passage from the population to the singular, as though stability of rates implied inevitability in every individual act. This does not prohibit predictive uses. It merely requires a distinction between a population rate and a judgement about an individual. A rate describes a regime of occurrences in a delimited population; the individual case requires additional criteria of membership and relevance. Without such criteria, the move from 'the many' to 'this one' becomes a logical abuse.[1]

The frequentist definition functions as conceptual hygiene. By saying that probability belongs to series, Venn restricts what the word can legitimately support. It describes mass regularities and therefore does not decide freedom, guilt, intention or necessity in individual cases. Probability so understood is not a sentence passed on an agent. It is a discipline governing how we speak about the world when only the population level is stable, given an explicit delimitation of comparable cases. Within that discipline,

[1]The Logic of Chance is the work's original English title.

a minimal vocabulary is imposed. A random experiment is defined as a repeatable procedure with sample space Ω. An event $A \subseteq \Omega$ is defined. On each repetition, A either occurs or does not. The relative frequency after n repetitions is written

$$f_n(A) = \frac{1}{n}\sum_{i=1}^{n} 1_A(\omega_i).$$

The frequentist thesis, in its idealised form, is stated as $P(A) = \lim_{n \to \infty} f_n(A)$, when that limit exists under fixed conditions. The gain is public clarity. The cost is that the limit lies beyond what we observe.

Historically, the motivation was also reactive. In the nineteenth century, Siméon Denis Poisson (1781–1840) distinguished objective and subjective probabilities, and a succession of authors deepened the unease with indifference as a foundation. John Stuart Mill (1806–1873) and Antoine Augustin Cournot (1801–1877) belonged to an intellectual environment in which 'chance' had to be connected to observable regularities. The support of George Boole (1815–1864) and Joseph Bertrand (1822–1900) reinforced the criticism of the idea that mental symmetry by itself could fix a number. In this setting, frequentism emerges as a demand for accountability.

4.2 Kollektiv, limit and randomness[2]

Richard von Mises (1883–1953) sought to give frequentism a complete form. He wanted a definition that would be not merely an empirical attitude but a foundation. At the centre of his programme is the Kollektiv, or collective: an idealised sequence of outcomes satisfying two requirements. What matters here is not the sequence as an archive, but the mechanism as generator. The collective is intended to capture 'mass phenomena', that is, arrangements in which frequency stability is expected under indefinite repetition.

It requires convergence. The relative frequency of an event stabilises as the sequence grows, as though there were a value towards which it tends. It also requires a condition of randomness, defined by invariance of the frequency under a specified class of rules

[2]Kollektiv (or 'collective') is the fundamental concept of the frequentist interpretation of probability developed by the mathematician and philosopher Richard von Mises.

for selecting subsequences. The idea is to protect frequentism from a conceptual artefact. If someone could systematically choose terms from the sequence so as to alter the frequency, then the 'limit' would not be a property of the regularity attributed to the process but an artefact of selection. The decisive question is always the same: which class of selection rules is admissible? Von Mises formulates this as 'insensitivity to place selection'. Admissible selections must not alter the limiting value. The rule acts as an antidote to gambling systems. If a selection rule guaranteed certain victory, what had been called a regularity would merely be the exploitation of bias.

The proposal is elegant, but it pays a high price. By tying probability to a limit in an infinite sequence, frequentism gains purity and loses contact with the finite world in which inference takes place. Any finite sequence, however extreme, remains compatible with any limit. A thousand heads in succession still does not by itself refute a probability of one half, because the definition lives at infinity. The point is not that extreme events cannot occur. It is that the criterion, in this form, becomes difficult to falsify within the horizon in which science makes decisions from finite data. The objection is logical, not psychological. A definition located at infinity becomes immune to a finite counterexample. What science does, then, is not to 'verify the limit' but to work with approximations, tolerances and error regimes. Publicity shifts from the 'limit itself' to the manner in which an approximation is judged sufficient in finite data.

Von Mises attempts to escape through a form of methodological confidence. Behaviour tends to stabilise under specified conditions of repetition, reference class and mechanism. 'In general' should be read as 'for classes of problems in which these conditions are plausible'. Even so, the fissure remains. The randomness he requires depends on restricting what counts as an admissible selection. Here a cost appears that is both conceptual and political. To sustain objectivity, one must legislate about what counts as illegitimate manipulation. Without such legislation, 'randomness' becomes a devotional word.

When twentieth-century mathematics tries to make this precise, it reveals the nature of the shift. In refinements such as that of Alonzo Church (1903–1995), illegitimate selections are restricted to computable procedures. This does not provide a physical essence of chance. It provides a methodological convention about which manipulations are to be regarded as illegitimate. Von Mises's 'randomness' thus becomes an ideal of immunity to selection bias relative to a class of operations, rather than a metaphysical property of the world. Within this framework, a sequence is called random relative to its resistance to computable selections. The gain is that 'we see no pattern' is replaced by a formal criterion of resistance to systematic methods. The cost is that the very notion of randomness becomes indexed to a class of procedures.

Frequentism learns to speak of randomness without reducing it to 'we see no pattern'. But to sustain objectivity it must legislate about what counts as legitimate manipulation. That legislation is not an appendix. It is the heart of the frequentist methodological pledge.

4.3 Reference class and finitude

Hans Reichenbach (1891–1953) recognises the sensitive point. If infinity is indispensable to the definition but inaccessible in practice, what are we doing when we speak of probability in science? Reichenbach's answer is not an ontology. It is a disciplined pragmatism.

He proposes that probabilities be understood as posits: this means proceeding as though there were a long-run frequency that would stabilise, even when only finite data are available. The regulative posit is not a belief in an invisible entity. It is a working rule that allows methods to be compared, uncertainties estimated and decisions guided under partial information. It does not assert a metaphysical truth that can be directly verified. It defines a practical regime of inference with public criteria of evaluation. Operationally, this means accepting that science works with sufficient convergence rather than absolute convergence. It also means accepting that 'randomness' can be taken in a weaker sense, as the practical inability to construct a selection rule that increases, on average, the number of correct predictions beyond what the frequency regime permits.[3]

This move saves frequentism from paralysis. At the same time, it shifts the problem to a point frequentism cannot avoid. If probability is frequency in the relevant series, everything depends on which series is chosen as relevant. Everything comes to depend on the chosen reference class. Here frequentist objectivity reveals its condition. It does not disappear, but becomes conditional and depends on public criteria of delimitation, measurement and stability of the mechanism.

[3] Posits (that is, regulative postulates) are statements or assumptions about future probabilities made on the basis of past frequencies, even when there is no guarantee that those frequencies will recur.

Reichenbach tries to discipline this choice through a requirement of homogeneity. The class should be broad enough to sustain a stable frequency and, at the same time, as homogeneous as the problem permits: the greatest justifiable number of cases similar to the case under consideration is included, and partitions are refined when statistically relevant properties alter the frequency. The tension between breadth and specificity does not disappear; it becomes an explicit part of choosing the reference class.

Here enters the problem that cannot be resolved by definition. There is never a guarantee that all relevant properties have been considered. Reichenbach therefore accepts that a statement about a single case is not uttered as a truth but as a weighting, almost as a wager, whose weight derives from the frequency in the class. This concession does not weaken the project; it makes explicit the public condition of the method. The weight is not a private conviction. It is a number tied to a publicly defended delimitation.

Finally, Reichenbach articulates an epistemological framework. He distinguishes primitive knowledge, in which prior probabilities are unavailable, from advanced knowledge, in which initial probabilities fixed by experience and frequencies are available. In the former, one makes 'blind weightings' guided by an inductive rule; in the latter, one calculates evaluated weightings. Bayes's rule enters as an instrument of confirmation, with the proviso that initial probabilities must be anchored in frequencies. In this respect Reichenbach can be described as an objective Bayesian without turning frequentism into a psychology of belief.

4.4 The single case and translation into a series

If the problem of the classical interpretation was the arbitrariness of the measure, the problem of frequentism appears as the arbitrariness of the class. Frequentism is weakest where the world demands a judgement about the singular, because its objectivity is, by definition, a property of the population level. This is a diagnosis, not an attack. A frequentist need not deny that singular events exist. In frequentism, probability is not taken to be a property of the singular as singular. It is assigned to classes of repetition. The singular enters through identification of a reference class and translation by modelling. The point can be stated without metaphor. The sentence '$P(A) = p$' acquires frequentist content only when one specifies 'in which series', 'under which conditions' and 'with what rule of repetition'.

The friction appears as a dilemma. Broad classes tend to be stable but insufficiently specific to the present case. Narrow classes tend to be specific but unstable, because few data support the number and variation dominates. The world presents singularities before it presents long series. The decision comes before the limit. This dilemma is the structural cost of publicity. The more public the delimitation, the more it must be defended by criteria that others can inspect; the more specific the delimitation, the fewer data support it.

The example of rain exposes this immediately. To ask 'what is the probability of rain tomorrow?' is, for the frequentist, to ask for a class of days sufficiently similar to tomorrow to be identified. The difficulty lies in 'similar'. Similar in which variables, with what tolerances, under which atmospheric model and using which historical base? The choice of delimitation is unavoidable. It is not given by the definition of probability, nor is it purely empirical, because it involves criteria of relevance and representation. Here an observation by William Feller (1906–1970) cuts sharply. Before speaking of the probability that the sun will rise tomorrow, one requires an idealised model defining what counts as a 'repetition' and as a 'selected world'. Without such agreement, the number has no place in the system.

The choice of delimitation is therefore an extra-probabilistic act of judgement. It lies at the threshold between theory and application. In this space, criteria of relevance, similarity and stability of the mechanism, none of which is supplied by the frequentist definition itself, must be brought to bear. This is the structural point. The reference class is constructed through criteria of relevance and modelling. Frequentism does not avoid this act of connection. It makes it the explicit location of judgement. The gain, again, is publicity. A critic need not accuse the analyst of 'belief'; the critic can challenge the delimitation, the assumed mechanism, the omitted variable or the protocol of repetition.

This is the point at which the criticism of the classical interpretation returns in another key. In the classical interpretation, the choice of space and measure entered beforehand. In frequentism, the choice of reference class enters beforehand. The form of the choice changes. The necessity of the choice remains. In both cases, interpretation enters before calculation. Frequentism merely moves its point of entry to a place where it can be audited.

4.5 Public error and repetition

Frequentism became decisive in science not only because of a philosophical definition but because it generated a technology of inference. Here Ronald Fisher (1890–1962), Jerzy Neyman (1894–1981) and Egon Pearson (1895–1980) become unavoidable names. Frequentism operates here chiefly as methodological technology. The focus is not to define what probability 'is', but to establish public criteria for the performance of procedures under repetition, given assumed conditions. Objectivity now takes an explicit form. It becomes a promise about error frequency. One does not promise 'truth' in each case; one promises control across repeated uses under declared conditions.

Fisher emphasises likelihood, estimation and tests as tools for extracting information from data under models. Even when he avoids committing himself to a direct semantics for probabilities about parameters, he retains the central requirement. Procedures should behave well under repetition. The focus shifts from the metaphysics of chance to the design of methods with comparable properties. He also criticises the loose metaphor of 'repeated sampling from the same population' when it becomes a verbal comfort. What matters is the design that guarantees comparability and the discipline of declaring assumptions.

Neyman and Pearson take this towards an almost institutional form. They provide a grammar of decision under controlled risk. One speaks of significance level and power, and of Type I and Type II errors. None of this is merely technical. It is a public but conditional promise. If a procedure is used repeatedly under the conditions of the design and assumed model, it will err with a bounded frequency. The guarantee concerns not the truth of a specific conclusion but the long-run reliability of the method. Here an ethic of public control takes shape. Inference becomes auditable, criticisable and reproducible in the methodological sense, even when individual results vary. Within this framework, the significance level sets an upper bound on the false-positive rate and power quantifies the ability to detect effects under specified alternatives. Inference becomes a policy of error, not a psychology of belief.

It is therefore important not to merge Fisher and Neyman–Pearson into a single philosophy of inference. A Fisherian significance test can be used to assess the incompatibility of data with a reference model and, in its classical formulation, does not require a complete decision rule between hypotheses. The Neyman–Pearson approach, by contrast, organises choice procedures through Type I and Type II errors and power, evaluated under repetition. Both belong to the history of frequentist statistics, but they answer different methodological questions.

Here lies frequentism's democratic contribution. It replaces the authority of intuition or individual belief with methodological accountability. Inference becomes an auditable process in which disagreements can be framed as criticisms of assumptions, designs, regimes of use and performance properties, rather than as contests of personal authority. It is also here that public misunderstandings arise. A p-value is not the probability that a hypothesis is true. A confidence interval is not a degree of belief that the parameter 'lies inside it'. They are guarantees of behaviour under repetition, and they exist only under the declared delimitation and model. When this conditional character is forgotten, the number begins to sell 'certainty'.

This is a rare virtue. It allows science to be a collective practice without requiring metaphysical unanimity about the ultimate meaning of probability. One can disagree about what probability 'is' and still agree about how a test should control errors and how an interval should achieve its coverage across repeated uses under explicit conditions. The Jeffreys–Lindley paradox, in particular, shows that different interpretations can lead to different judgements about 'significance' from the same data. This is not a defect of arithmetic. It is a reminder that semantics and the decision rule matter, and that 'publicity' is not the same as 'univocity'.

4.6 Replication and delimitation

In replicable trials, frequentism is particularly powerful: planned experiments, clinical protocols, industrial quality control and repeatable sampling. In such contexts, repetition is not a fiction but part of the design. The frequentist requirement becomes natural. Error control, coverage and power all correspond to commitments that communities can test and compare. Here the interpretation fits the practice. By design, the world supplies series. The method promises performance in series. Publicity becomes concrete because others can repeat, audit and compare.

Frequentism shows its best face here. It does not ask us to trust an agent's belief. It asks us to trust a procedure that anyone else can audit. When science is a collective enterprise, this requirement has political and epistemic value. It protects against the urge to conclude too soon and creates standards of responsibility that outlast the enthusiasm of the analyst. That protection, however, is not automatic. The delimitations must be published. The protocols must be published. The choices defining 'same population', 'same condition' and 'same mechanism' must be published. Without this, replication becomes a dispute over language.

In singular predictions, frequentism operates by mapping the case onto a reference class constructed through criteria of similarity and modelling. Probability is then the rate in that class under the assumed regime. Individual risk, failure of a unique component, storm forecasting, diagnosis in a particular patient: in every such case, frequentist probability depends on how the series of similar cases is defined. Good models and good delimitations can be constructed, and then the mapping works. The delimitation can also be disputed, and the probability then becomes disputable. The calculus remains stable. The delimitation is not unique. The decisive point is that frequentism does not claim to abolish judgement. It seeks to move judgement to a place where it can be criticised without psychologising it. Anyone who disagrees with the number should be able to say precisely what changed in the delimitation.

This shows clearly what frequentism actually offers. It does not eliminate judgement. It disciplines judgement through public rules and long-run guarantees. When the singular is unavoidable, it requires the singular to be incorporated into a practice of comparison. The publicity here is the publicity of error, not the publicity of certainty.

4.7 Synthesis

Frequentism attempts to escape classical circularity by shifting the point of support. It exchanges symmetry for regularity, indifference for series and equipossibility for frequency. In doing so, it produces virtues that have marked modern science. It provides a public form of justification, a language of audit and a technology of error control that turns inference into a governable practice. Its normative core can be stated as follows. Given an explicit and repeatable delimitation, procedures should have controllable performance under repetition. Frequentist objectivity is not the claim to speak 'from nowhere'. It is the claim to speak from a declared delimitation.

The price of this virtue appears in three principal costs, though they need not be forced into a rigid scheme. There is dependence on long-run idealisations, because stability is conceived on the horizon of extended series. There is dependence on the reference class, because objectivity requires selection of the relevant delimitation. And there is the structural difficulty of the single case, because for the frequentist probability belongs to the population level and the singular enters through mapping. This third cost is not accidental. It is a direct consequence of the promise of publicity. For the frequentist, probability has its proper force where there are rules of repetition and comparison.

Frequentism does not solve the interpretative problem. It reconfigures it. It shows that an important part of the meaning of probability is methodological, connected with public control and the reliability of procedures. In doing so, it opens two principal directions. One emphasises degrees of belief and norms of updating. The other emphasises objective chance in the world, often articulated as a propensity or disposition of physical set-ups. In both cases, the aim is to supply what 'frequency' and 'method control' do not exhaust. The history of the terminology also records the division. Maurice George Kendall (1907–1983) contrasts frequentists and non-frequentists when discussing the attempt to avoid opinion and anchor probability in objective properties of real or hypothetical populations. John Maynard Keynes (1883–1946) had already used 'frequency theory' as a chapter heading in 1921. The conflict was never merely mathematical. It was semantic, methodological and public.

What is learnt here is a pattern that the whole book will pursue. Each interpretation solves some problems at the cost of admitting others. Intellectual merit does not lie in denying the price. It lies in making the price explicit and deciding when it is acceptable. Frequentism paid the price of the reference class in order to secure the publicity of method. The next interpretation, propensity theory, will attempt the opposite route. It will pay the price of a more robust ontology, with objective chance as a causal disposition of the set-up, in order to confront the problem of the single case directly and offer an objectivity inscribed not merely in procedure but in the physical structure of the world itself.

5 Propensity Interpretation

(When the world seems to demand more than series)

The previous chapter showed why frequentism is attractive as a project of publicity. It promises objectivity by relocating probability into auditable practices: series, protocols, populations and replications. That shift has real advantages. It also has a cost that is not peripheral. When a decision concerns a singular case, frequentism tends to require a compulsory translation. The unique event must be rewritten as a case belonging to a reference class. The translation may be correct and still be insufficient for the way in which the world demands action.

It is at this point that propensity theory takes shape. The wager is ontological. Probability is not merely observed regularity or only a discipline of method. It purports to denote an objective chance associated with a causal set-up, not an agent's degree of belief and not a modelling device treated as neutral. This move exacts a specific conceptual price. It requires a metaphysics of dispositions and forces a negotiation between two structures that do not fit together automatically. On one side is the formal symmetry of the calculus, which operates on propositions and carries no temporal direction. On the other is the intrinsic asymmetry of causality, which points from the set-up to the outcome and whose direction is not reversed by algebraic manipulation.

If chance is objective, it is not merely a description of the world. When known, it also constrains rational judgement. This does not amount to turning ontology into psychology. It simply recognises that a theory of objective chances which does not explain how they are co-ordinated with rational degrees of belief is left with a strong ontology and an improvised epistemology. The normative bridge must be stated.[1]

A modal distinction guides the entire chapter. Frequency describes what happened, the record of what occurred. Propensity purports to describe what would tend to happen, as a disposition of the set-up prior to any realisation. Frequency is a retrospective portrait. Propensity aims to be a prospective description.

[1] Lewis's Principal Principle. It formalises the intuition that, when an objective chance is known and the additional evidence is admissible, rational degree of belief should equal the value of the chance.

5.1 Quantum motivation

Karl Popper (1902–1994) brings probability close to a vocabulary that became central in twentieth-century physics. There are processes in which reducing uncertainty to mere ignorance of hidden details does not satisfy the realist impulse. In such cases, probability does not sound like a mere description of epistemic limits. It sounds like part of the phenomenon itself, something belonging to the way in which the system can produce outcomes.

Two examples are enough to fix the intuition without turning this chapter into a treatise on physics. In radioactive decay, an individual atom is assigned a chance of decaying within a certain interval even when nothing in its macroscopic state 'announces' the instant. In interference experiments with particles sent one at a time, a pattern emerges as a statistical regularity, yet each individual occurrence still appears to require a description that cannot simply be reduced to 'ignorance of microdetails'. The propensity appeal here is to an objective chance connected with the generating conditions of the experiment.

The move is also strategic. Popper saw in some orthodox readings of quantum mechanics the danger of a disguised subjectivism, especially when the narrative of collapse appeared to depend on the observer's epistemic position. In this setting, propensity theory is an attempt to sustain a realist reading. Probabilities do not reside in the agent's knowledge. They are objective features of the experimental system.

There is also a philosophical precursor worth naming because it changes the historical framing of the problem. Charles Sanders Peirce (1839–1914) had already worked with the idea of tendencies and habits belonging to types of situations, anticipating part of the motivation behind propensity theory. This does not displace Popper. It merely shows that the friction between the single case and publicity is recurrent rather than episodic.[2]

[2]Peirce and habit. Peirce's notion of 'habit', connected with tendencies and regularities of a type of situation, anticipates part of the idea of objective disposition, although contemporary debate has refined it in the language of chances and mechanisms.

5.2 Disposition of the experimental set-up

The most defensible form of propensity theory does not treat a propensity as an essence lodged in an isolated object, as though every coin silently carried a number inscribed within it. It treats propensity as a relational disposition: a type of preparation, a type of test, a type of interactive environment. The propensity does not inhabit 'the coin' in a vacuum. It emerges from how the coin is tossed, how it is caught and the physical regularities stabilised by the procedure. In general terms, it is a property of the experimental set-up understood as a causal configuration that can be reproduced as a type, even when the event in question is singular.

To reject essentialism precisely, the point should be stated without metaphor. A propensity is not a monadic property of an object, as though it were an attribute carried in isolation. It is a systemic property: the disposition of a set-up to produce a certain outcome under specified conditions. It belongs to the complete experimental situation, including preparation conditions, material constraints and stabilising mechanisms.

This position earns the right to speak about the single case, but pays through a requirement that must be declared if it is not to become an excessive promise. The reference-class problem does not disappear. It returns as a problem of individuating the set-up. If 'the same type of causal configuration' is not to become an elastic label, a public criterion is needed for what counts as relevant in the description of the generating conditions. The theory shifts the problem. It does not erase it.

The analogy with dispositions helps, provided that it is used carefully. Fragility is not the mystical destiny of a glass. It is a disposition manifested under certain triggers. Likewise, propensity is a tendency towards certain outcomes under conditions of preparation and test. This emphasis blocks an essentialist reading. Propensity is not a probabilistic soul of the object. It is a dispositional feature of the system in operation.

5.3 The metaphysics of probabilistic dispositions

The metaphysical cost of propensity theory is not a detail. It is the core of its ontological commitment. In contemporary metaphysics, dispositions have often been analysed through counterfactual conditionals of the form ‘if it were in conditions C, it would manifest R’. This scheme is intuitive for dispositions such as fragility, solubility and flammability. It becomes less tractable when the disposition is probabilistic.

The reason is simple. Propensities are not binary dispositions. They come in degrees. A chance of 0.3 cannot be reduced to ‘if the set-up were instantiated, then it would occur’, because the very form ‘would occur’ is binary. Trying to translate it as ‘it would occur 30% of the time’ already reintroduces frequencies as the explanation, precisely what propensity theory sought not to take as a definition. A propensity theorist may accept frequencies as evidence, but loses the target if frequencies are accepted as a reductive analysis of the content of propensity.

Here a metaphysical bifurcation opens and should be presented honestly.

One route is strongly realist. Propensities would be irreducible causal powers, genuine properties of the set-up with no need for reduction to some further basis. This route preserves the quantum motivation and accepts that some chances are brute facts of the world, just as certain constants and laws are treated as brute facts in many scientific ontologies.

The other route is reductive or, at least, parsimonious. Propensities would be complex counterfactual patterns sustained by underlying mechanisms, so that the fundamental ontological content is not the chances themselves but physical structures which, under type-level repetition, make the stability of frequencies intelligible. This route has appeal in macroscopic domains, where the ‘random’ may be suspected of arising from complexity, chaos and sensitivity to initial conditions.

The debate is well known and should not be reduced to a single metaphysics of powers. Nancy Cartwright (1944–) works with capacities and local mechanisms, emphasising that causal tendencies operate in concrete contexts. Stephen Mumford (1967–) defends dispositions as powers and criticises simple reductive analyses. Alexander Bird (1964–) articulates a dispositionalist metaphysics in which properties and laws receive their own treatment. These approaches converge in rejecting a purely nominal reading of disposition, but diverge over the status of laws, powers and conditions of manifestation. For propensity theory, the relevant lesson is that calling a chance a disposition does not complete the analysis: one must still explain what kind of probabilistic property is being postulated and how it can be identified empirically.

5.4 Frequency as evidence

Propensity theory does not abandon frequency. It repositions it. Frequency becomes evidence rather than definition. It is the public trace through which we infer something that, within the proposed ontology, is prior to that trace: the causal disposition of the set-up.

The distinction between actual and virtual repetitions makes this role explicit. Frequencies observed in actual sequences function as an indirect test of a conjecture about what would happen under indefinite repetitions of the same type of generating conditions. A set-up may be singular in its execution. It must nevertheless be typifiable in principle, if the attribution of a propensity is not to become an untestable metaphor.

A conceptual diagram, described in words, helps to fix the architecture of levels without requiring a figure. Three planes can be distinguished. On the ontological plane there is a chain from the experimental set-up to propensity as the objective chance of that set-up and thence to possible manifestations in singular events. On the methodological plane there is another chain, from repetition of the type of set-up to observed frequencies and statistical summaries and thence to inference about the propensity as an ontological hypothesis. On the epistemic plane there is the normative circuit, from an initial degree of belief or prior probability over hypotheses to updating by evidence, with co-ordination with objective chances when these are known or assumed.

To prevent 'probability' from concealing ambiguity, it is useful to fix the notation so that the levels remain visible. At the model level, we work with a probability space $(\Omega, \mathcal{F}, P)$, where P is the model measure. At the ontological level, when the context requires realist emphasis, let the objective chance of the set-up be denoted by Ch($\cdot$). At the epistemic level, let π(H) denote the initial probability over hypotheses and π(H | E) the degree of belief updated by evidence.

Let H be a hypothesis about the set-up, understood as a description of the mechanism and its chances. Let E be the available evidence, for example a pattern of frequencies observed in repetitions of the type of set-up. Updating can be expressed as

$$\pi(H \mid E) = \frac{P(E \mid H)\, \pi(H)}{P(E)}.$$

The delicate point is this. P(E | H) is probability within the model. It can be connected with the chances Ch described by H, provided that the model–world link is made explicit. By contrast, π(H) and π(H | E) are epistemic quantities, probabilities over hypotheses. They are not, in themselves, propensities.

A simple example fixes objectively the difference between direct and inverse probability. Consider a set-up A for tossing a coin. The propensity thesis would say that the set-up has an objective chance Ch_A(heads) = x. This is an ontological assertion about the set-up. Now suppose that a family of hypotheses H_x encodes 'the set-up has chance x'. After observing data E, one updates π(H_x | E). This is inference about hypotheses. It is not a propensity that 'runs from E to H'. It is a degree of evidential support.

From this point onwards, one sentence must prevent a category mistake. Reversing an equation is not reversing causality. Propensity, as a causal disposition, is directional, from set-up to outcome. Inverse probability organises reasoning in the light of indications. It does not reverse the arrow of the world.

5.5 Causal asymmetry and Bayes

Paul Humphreys (1950–2022) crystallises the tension in an argument that has become a standard reference point. If propensities are causal dispositions, they carry an asymmetry from set-up to outcome. Yet probability calculus allows systematic passage to inverse conditionals and does not preserve causal direction as part of its content. It organises relations between propositions under general rules. Causal direction is not an algebraic property.

The problem is one of category. Probabilistic relations between propositions, even when they concern chances, are not by themselves causal relations in the world. To confuse the levels is to impose an excessive ambition: that the structure of the calculus should faithfully mirror the structure of dispositions.

The propensity theorist is pressed to choose an exit, and neither option is cost-free. If every probability is insisted upon as a propensity, the difference between causal direction and inferential direction tends to disappear, at the price of conceptual obscurity. If probabilities over hypotheses in inferential use are conceded to be epistemic, the distinction is preserved, but 'probability' is acknowledged to name different things, some ontological and others cognitive.

There is a reply that does not deny Humphreys's argument but limits its scope. Instead of claiming a simple identity, it distinguishes levels. Propensities belong to the ontological level of the set-up. Inverse conditionals belong to the epistemic level of evidential support. Thus π(H | E) does not represent a 'reverse propensity'. It represents the degree to which the evidence supports the hypothesis that a certain direct propensity is present in the set-up.

This reply preserves the realist core of propensity theory while abandoning the ambition that every probability in every use should be one and the same entity in the world. On this reading, Humphreys's criticism does not destroy the notion of objective chance. It destroys the strong identity claim.

5.6 Co-ordinating chance and degree of belief

David Lewis (1941–2001) provides the standard formulation of the normative bridge between chance and degree of belief. The idea is this. If the chance of an event is known at time t, and if the additional evidence E is admissible, then the rational degree of belief in A should align numerically with that chance.

A clean way of expressing this is to write, with Cr for rational degree of belief and Ch_t(A) for chance at time t,

$$Cr(A \mid Ch_t(A) = x \ \wedge \ E) = x.$$

The value of the principle here is not rhetorical. It makes explicit a co-ordination that often remains tacit. If one says that probability is objective chance while also using probabilities in inference and decision, one must explain why objective chance should guide rational judgement.

The delicate point is the so-called admissibility problem. What kind of information may be added to 'the chance is x' without defeating the directive? If E includes information about the event's own future, the co-ordination dissolves. The same occurs if E includes improperly correlated information.

This difficulty is not accidental. It mirrors the arrow of time: a chance at time t is intended to summarise, from the state and laws up to t, what may occur after t. 'Admissible' information therefore tends to be information that does not import content about the event's own future into the conditioning. When E carries features of the outcome, the co-ordination breaks down not because of a failure in the calculus but because the temporal asymmetry that also characterises causality has been violated.

Later refinements seek to preserve the core of the principle without rendering it circular or empty. One influential proposal is associated with Ned Hall (1966–), who reformulates the bridge in terms of expected chance given the past.[3]

This debate reveals something directly relevant to the chapter. Chances are temporal: they evolve with the state of the world and with laws governing transitions. Evidential support is logical and comparative between propositions. The normative bridge is where these two structures meet, which is why the problem is not a technical detail. It is a point of contact between ontology and rationality.

The Principal Principle does not, by itself, constitute a complete theory of the nature of chance. Its function is normative: to make explicit how rational credences should respond to information about chances when the remaining evidence is admissible. The metaphysical analysis of chance remains a further question, and it was precisely at this point that Lewis later developed his best-system proposal (Lewis, 1980; Lewis, 1994).

[3]The admissibility problem. 'Admissible evidence' is evidence compatible with applying the Principal Principle. Hall's Big Chance Principle reformulates the idea using expected chance conditional on suitable past information, excluding future information that would trivialise the principle.

5.7 Contemporary variants and objections

Propensity theory is not monolithic, and this diversity is part of the philosophical diagnosis. Some versions preserve the intuition of causal disposition while retreating from single-case chances.

Broadly speaking, Ronald Giere (1938–2022) defends a version in which propensities make more sense as dispositions to produce stable frequencies under repetition than as fully numerical chances assigned to a single event. The gain is a reduction in the suspicion of an 'idle property', a property that would exist before any manifestation and might, in principle, never leave a trace. The cost is a return towards the spectre of the reference class, because stability again depends on the form of type-level repetition.

There are also more ecumenical versions in which 'propensity' names probabilistic causal tendencies, while 'probability' remains a formal tool of inference. This line is compatible with the move suggested by Wesley Salmon (1925–2001), who emphasises the causal role of propensities in explanation. The gain is to avoid a strong identity between chance and standard probability. The cost is to accept semantic plurality in the term 'probability', acknowledging that it operates across distinct families.

The literature also discusses whether propensities require a physical categorical basis or may instead be fundamental properties. In some quantum contexts, certain formalisms use objects that behave like 'quasi-probabilities', sometimes taking negative values. These should not be confused with objective chance in the sense used in this chapter. Objective chance here remains confined to the interval [0, 1]. The purpose of those formalisms is different, concerning representations and reconstructions of states.[4]

Finally, some approaches bring the debate closer to contemporary theories of causation and explanation. Michael Strevens (1969–) emphasises mechanisms and stable explanatory capacities even where the assignment of a numerical chance is not ontologically central. In another strand, James Woodward (1956–) and Judea Pearl (1936–) provide structural languages for rigorously separating causal relations from merely probabilistic relations. For this chapter, the value of these approaches is didactic. They show why

[4]Quasi-probabilities. In some quantum representations, mathematical objects arise that behave analogously to 'probabilities' and may take negative values. This should not be confused with objective chance. Here, objective chance remains a concept in the interval [0, 1], associated with generating conditions and propensities in the sense of this chapter.

the move from probability to causality requires additional structure and why causal asymmetry cannot be automatically 'read off' from the formal symmetry of the calculus.

A recurrent objection here is whether propensities require fundamental indeterminism. Part of the quantum motivation suggests that they do, at least in certain domains. There is, however, a compatibilist line: in deterministic but chaotic systems, practical unpredictability and sensitivity to initial conditions may sustain stable assignments of 'chances' at the macroscopic level, understood as dispositional descriptions of the set-up under unavoidable perturbations. On this reading, propensity does not prove indeterminism; it organises, in modal language, the role of type-level habits and regularities, approaching the Peircean background without collapsing into mere frequency.

5.8 Best-system interpretations of chance

Propensity theory is not the only contemporary attempt to preserve objective chances. In best-system interpretations, developed from the work of David Lewis, chance is analysed in the context of laws of nature. The central idea is Humean: one begins with the total history of the world and asks which true system organises it most efficiently, balancing simplicity, informativeness and, when the system is probabilistic, fit to the observed history. The probabilities appearing in the laws of the winning system then play the role of objective chances (Lewis, 1994; Hoefer, 2019).

The difference from frequentism is important. Observed frequencies contribute to the fit of a probabilistic system, but they do not by themselves define chance. The difference from propensity theory is also substantive. Chance need not be a causal power localised in a device or set-up; it may supervene on the global pattern described by the best system. Objectivity is therefore located less in a local disposition than in the best systematisation of the world's regularities and laws.

The advantage of this strategy is that it avoids part of the metaphysics of powers without reducing chance to mere frequency. The cost appears in the criteria of the system itself. Simplicity, strength and fit must be compared, and there is no completely neutral metric that resolves every dispute about which systematisation is best. Moreover, different Humean versions diverge over determinism, non-trivial chances and temporal dependence. The theory therefore does not eliminate choices; it moves them to the evaluation of the systematic virtues of laws.

For the pluralism developed in this book, the lesson is direct. There is more than one way to locate objective chance: in a disposition of the set-up, as in propensity theory, or in the probabilistic structure of the best system of laws. These positions are not equivalent and should not be grouped together merely because both speak of chances. The contrast reinforces the need to state where objectivity is being located and what metaphysical or methodological price accompanies that location.

5.9 Domains of application and limits

Concrete cases help to delimit the scope of propensity theory without losing sight of its central thesis.

In the quantum domain, the realist motivation is strong. A propensity theorist can maintain that there are objective chances associated with generating conditions and that these chances are part of the world, not merely a summary of ignorance. Here the thesis of an irreducible probabilistic disposition appears more natural, though also more ontologically demanding.

In macroscopic domains, the situation is more ambiguous. When one says 'a 60% chance of rain tomorrow', many meteorological practices operate through statistical modelling based on populations of similar situations, updated by data. This appears less like a singular causal disposition and more like an initial probabilistic assignment disciplined by models and historical frequencies in reference classes.

In personalised medicine, the risk to this patient oscillates between two readings. It may be treated as a statistical extrapolation from populations and subpopulations. It may also be interpreted as an attempt to capture dispositions of the organism under intervention, partly causal and partly inferential. The chapter need not deliver a universal verdict. It must state the cost of each way of locating objectivity.

This is the point of the debate. Frequentism locates objectivity in the long-run behaviour of populations and procedures. Propensity theory attempts to locate objectivity in the world, as a causal structure of the set-up, thereby preserving the single case. The gain is the recovery of the sense of 'objective chance'. The cost is both ontological and methodological: ontological because it requires a metaphysics of probabilistic dispositions, methodological because it requires public individuation of the set-up and explicit co-ordination between chance and degree of belief.

This prepares, without anticipating too much, the questions that open the next chapter. If propensities are accessible only through frequencies and modelling, must we really treat them as ontological entities, or is it enough to treat them as explanatory devices? If the link between chance and degree of belief depends on a normative principle, does that principle not reintroduce, by another route, the need for initial probabilities and, in particular, the dispute over what counts as 'subjective' and what counts as 'rationally required' in the choice of $\pi(H)$? If inference about hypotheses operates with $\pi(H \mid E)$, in what sense does that probability remain public, and in what sense does it remain dependent on unavoidable initial choices?

5.10 Synthesis

Propensity theory arises in response to the point at which frequentism, despite its public virtue, begins to require a structurally burdensome translation. When a decision concerns a singular case, the frequentist method asks that the event be rewritten as a case belonging to a reference class. That may be correct and still be insufficient for the way in which the world demands action. The propensity proposal is to reverse the location of objectivity. Instead of locating it only in the performance of procedures across series, it seeks to locate it in the causal set-up that produces the event.

This shift has an evident gain. It recovers the sense of an objective chance prior to the statistical trace and connected with the generating conditions of the experiment or process. The single case therefore ceases to be merely a 'point in the series' and becomes treatable as a manifestation of a disposition of the set-up, even when no long sequence is immediately available to support it. The price, however, is equally clear. Propensity theory assumes an ontological commitment: it must say what a probabilistic disposition is, in what sense it is real and how it is anchored in the world.

The chapter has also established a discipline of levels. Frequencies do not disappear. Their status changes: they cease to be a definition and become evidence. It is through actual repetitions, when available, and typifications of the set-up, when necessary, that the hypothesis of a propensity is indirectly tested. The problem of delimitation is therefore not abolished: it returns as the problem of individuating the set-up and of providing public criteria for what counts as the 'same type' of generating conditions. Publicity here lies not in 'avoiding choices' but in exposing them.

Best-system interpretations add a distinct alternative: they preserve the language of objective chance without identifying chance with a local power of the set-up. In exchange, they make decisive the choice of the best systematisation of laws, according to criteria of simplicity, strength and fit. The map of objective chance therefore comes to include at least two important contemporary families, propensities and best systems, which share realism about chance without sharing the same ontology.

Finally, the deepest tension was asymmetry. Causality has a direction: from set-up to outcome. Probability calculus by itself carries no such direction and permits inverse conditionals without causal commitment. Humphreys's argument delimits the scope of propensity realism: chances may be ontological; probabilities over hypotheses may be epistemic; and reversing an equation is not reversing the arrow of the world. If propensity theory is to be more than realist rhetoric, it must explicitly state the normative bridge between chance and degree of belief under conditions of admissibility.

Between the ontology of propensities and the psychology of degrees of belief, the logical interpretation appears as a third attempt at objectivity. Instead of locating probability in series, as frequentism does, or in causal dispositions of a set-up, as propensity theory does, it locates probability in the logical relation between a hypothesis and a body of evidence, as a measure of how conclusive an argument is. The gain is a promise of objectivity without infinite series or a robust metaphysics of physical dispositions. The cost is to shift the discussion to how the body of evidence is fixed and to the language in which hypotheses and evidence are formulated, thereby reintroducing, by another route, the problem of delimitation and publicity.

What is learnt, then, is a reconfiguration of the same pattern running through the book. Frequentism purchases publicity through error control in series and discipline in delimitation. Propensity theory purchases the single case through an ontology of dispositions and an explicit co-ordination between world, model and rational judgement. The logical interpretation will attempt to purchase objectivity through evidential semantics, that is, through the idea that 'probability' measures rational support between propositions. The merit lies not in denying the cost, but in making it visible and deciding when it is acceptable.

6 Logical Interpretation

(Confirmation, language, and the cost of making evidential support public)

The previous chapter ended with a problem of co-ordination between chance and degree of belief. If objective chances exist in the world, how should a finite mind adjust its degrees of belief in the face of partial evidence without falling into arbitrariness? Logicism enters this scene as a deliberately intellectualist wager. It gives priority to the logical structure of thought and justification over causal processes in the world or psychological habits of the mind. Perhaps probability, in its inferential use, need not be anchored in frequencies or causal dispositions. Perhaps it can be understood as an impersonal relation between propositions, a degree of confirmation that measures how strongly evidence supports a hypothesis, given a body of information and a form of description.

The ambition is this: to turn 'degree of belief' into something public without reducing it to psychology. Logicist probability aims to be intersubjective because it is normative. It is meant to hold for any rational agent who accepts the same descriptions, the same linguistic commitments and the same standards of inference.

This explains why logicism appears, at once, to be the most elegant answer to the co-ordination dilemma and one of the most difficult programmes to sustain. It promises validity without appealing to physical mechanisms and without depending on a population. In return, it requires a demanding formal infrastructure, a strong notion of total evidence and a robust defence of what counts as 'rationality' in induction. The risk is already present at the starting point. Impersonality does not come from the world as an experimental set-up. It comes from rules of description. And rules of description have histories, choices and frictions.

Chapter 3 already established a thesis that now reappears on another plane. There is no cost-free neutrality in the passage from calculation to the world. In the classical interpretation, the friction appears as dependence on measure and parametrisation. Here, it appears as dependence on language and inductive rules. The movement is the same; only the stage changes. What was previously a 'uniform measure' in a continuous space becomes 'descriptive symmetry' in the space of hypotheses. The question, then, is not merely 'what does the evidence say?'. It is also 'in what language does the evidence speak?' and 'under what inferential constitution is the evidence entitled to support generalisations?'.

6.1 Antecedents of probabilistic logicism

Before logicism acquired its mature form in the twentieth century, Victorian debates had already rehearsed a central idea. Induction would not be merely a psychological habit or a matter of practical prudence. It would have a rational structure, and that structure could be treated with a discipline akin to that of logic.

William Stanley Jevons (1835–1882) is an early landmark of this impulse. His 'inductive logic' attempts to quantify the weight of evidence, bringing inductive justification closer to a controllable calculus. The point here is not the technical correctness of each proposal, but the philosophical gesture. Confirmation ceases to be merely a feeling of conviction. It begins to aspire to a public status, in the sense of being discussable and criticisable by rules.

This attempt, however, arises already under tension, because the ideal of publicity requires class and form. And logicism, in seeking publicity outside the population, shifts the problem to description. This is why a figure associated with frequentism can serve as an anticipatory critic of logicism.

John Venn (1834–1923) insists that 'probability' does not stabilise without empirical support and without a reference class that makes the statement intelligible. Even when probability is treated as something rational, Venn forces an uncomfortable question: rational for whom, in which language, under which delimitation? This kind of objection does not destroy logicism. It compels it to recognise that its publicity is linguistic and normative, not physical.

Between Jevons and Keynes, another link matters because it connects indifference, symmetry and language.

William Ernest Johnson (1858–1931), known as W. E. Johnson, is an important link in the Cambridge tradition. Keynes explicitly acknowledged his influence, and Johnson's results on symmetry and sufficiency left traces in later developments of inductive logic. There is no need to assume a direct causal line from Johnson to every subsequent author: the historically secure point is that he significantly influenced the intellectual environment in which Keynes formulated his theory and anticipated ideas that would reappear, in different forms, in work on exchangeability and inductive methods.

This background helps to avoid a misleading reading. Logicism did not arise from nowhere with Keynes. It was the culmination of an effort to extend the discipline of logic to the uncertain domain of induction. The difference is that, with Keynes and Carnap, this effort acquired an explicit architecture.

6.2 The ideal of confirmation

John Maynard Keynes (1883–1946) occupies a distinctive place in this genealogy. For him, the probability relevant to science and decision is not primarily frequency or propensity, but a rational relation between premises and conclusion. Given evidence (E) and a hypothesis (H), there is a fact about how strongly E supports H. This fact does not depend on who judges, in the psychological sense. It depends on what a rational judgement ought to concede in the light of the evidence and the way in which that evidence structures the space of alternatives.

Here it is worth tying this point to the distinction that runs through the book between ontology and epistemology. In frequentism and propensity theory, probability tends to be treated as something that belongs to the world, whether as a limit of frequencies or as a tendency of the set-up. In logicism, probability belongs primarily to the argument. It measures inferential support. The world enters as content, as premise, as evidence. But the primary object is the relation between propositions that describe the world.

The Keynesian project also inherits, in a new key, one of the engines of the classical interpretation. When the evidence does not discriminate among mutually exclusive and exhaustive alternatives, symmetry in the description of the hypotheses, given the available language, warrants treating them as equipossible in the logical sense. 'Objectivity', then, does not arise from the world as mechanism. It arises from the structure of the description, that is, from the way the problem has been formulated and from what, within that formulation, counts as a discriminating reason. It is a principle of indifference recast in a logical-linguistic key, with an unavoidable cost. Impersonality becomes dependent on the language adopted and on its criteria of symmetry.

At this point, Bertrand's lesson reappears at the level of language.

Joseph Bertrand (1822–1900) showed that, in certain continuous problems, the 'answer' depends on how randomness is defined, that is, on how the space of possibilities is represented. Logicism does not exorcise this ghost. It internalises it. The problem of the 'uniform measure' becomes the problem of 'descriptive symmetry'. Objectivity does not escape convention. It migrates from geometry to grammar.

Keynes adds another decisive feature. Not every relation of support is totally comparable or measurable by a single number. There are cases in which the evidence appears to support one hypothesis more strongly than another without providing a universal scale. This refusal to reduce all inductive rationality to a single yardstick is both philosophical honesty and an operational obstacle. Applied science, particularly when decisions involve asymmetric costs, frequently requires metrics.

This point opens an important line of thought that preserves the Keynesian spirit while rejecting the demand for total measurement.

L. Jonathan Cohen (1923–2006) develops a theory of evidential support in which confirmation can be comparative without necessarily being numerical, and in which degrees of support can resist reduction to point probabilities. The gain is that it preserves a real feature of rational practice in settings of fragmentary evidence.

The cost is to reintroduce the problem of publicity in a new form. If there is no number, what guarantees auditability? The typical answer is a discipline of explicit comparison, with rules of relevance and inferential defeat. In this branch, logicism becomes less a calculus than a jurisprudence.

The conclusion of this section can be stated in the terms of the book. Logicism attempts to make evidential support public without reducing support to psychology and without anchoring probability in a population. To do so, it relocates objectivity to language and to norms of inference.

6.3 Confirmation and objective chance

Rudolf Carnap (1891–1970) systematises the logicist ideal through explicit conceptual engineering and an institutional ambition characteristic of logical empiricism. His theory of confirmation is not merely a technical component. It belongs to a broader intellectual project, connected with the ideal of the unity of science and with the idea that philosophical disputes can be dissolved through clarification of language, formal reconstruction and public criteria of acceptability.

Carnap's most operational distinction separates two uses that, in scientific practice, are easily conflated. On the one hand, there is probability as confirmation, a measure of the degree to which evidence confirms a hypothesis within a fixed language and set of inductive rules. This is the logicist core, often designated probability_1. It answers a question of this kind: given our evidence and our language, how rational is it to maintain H?

On the other hand, there is probability as frequency or objective chance, associated with observable regularities or with the chances of a physical set-up. This is the role of probability_2, close to frequentism and to the ontologies of chance discussed in propensity theory. It answers a different question: independently of what we know, what is the tendency or regularity of the phenomenon?

The distinction is operational. When one says 'the hypothesis is highly probable', one may be claiming that the hypothesis is well confirmed by the available data, or that the phenomenon has high probability_2 in the world. Carnap attempts to construct a rigorous theory for the former and, by design, retreats from the latter.

This retreat is part of the ideal of publicity. If we want a public notion of evidential support that is neither frequency nor objective chance, we must define it by explicit rules. And this requires language, criteria and an inductive policy that does not remain concealed. Publicity here does not mean 'the world speaks for itself'. It means 'the procedure of confirmation can be reconstructed and audited'.

At this point an element enters which, once made explicit, renders the programme simultaneously more intelligible and more vulnerable. Carnap defends methodological tolerance regarding the choice of formal languages. There is, by decree, no single correct language. There are languages that may be adopted under criteria. This connects directly with his allowance for a plurality of inductive rules. Confirmation is not a brute property of the world. It is defined under a language and an inductive constitution.

In The Continuum of Inductive Methods, Carnap makes this plurality explicit by constructing a family of inductive methods rather than postulating a single inevitable confirmation function. The parameter that regulates the relative weight assigned to initial information and observed evidence shows that Carnapian tolerance is not an absence of rules: it is a formally structured plurality whose choices still require criteria of adequacy to the problem (Carnap, 1952).

This background also helps to understand an indirect influence.

Ludwig Wittgenstein (1889–1951) is, for logical empiricism, a twofold reference point. In the Tractatus Logico-Philosophicus there appears the aspiration that logical form, when properly exposed, should provide clarity and dissolve confusions. In the Philosophical Investigations, by contrast, there is the reminder that meaning depends on use, practices and forms of life. The point useful to this book is the following: even when logicism formalises, it does not eliminate the fact that language has history and friction. The promise of impersonality, therefore, is never cost-free. It must be sustained by a policy of description accepted by a community.

6.4 Inductive schemes and evidence

The central tool of the Carnapian programme is the idea of a confirmation function, usually called a c-function. Conceptually, a c-function is a rule that, given a formal language, a body of evidence and a hypothesis formulated in that language, returns a degree of confirmation.[1]

A minimal example fixes the idea without turning it into a calculation. Imagine a language that describes individuals by means of a single predicate, something such as 'being F'. The evidence says that certain observed individuals are F and others are not. The hypothesis we wish to assess is a modest projection, for example, whether the next individual will be F. Two different c-functions may agree with the evidence already

[1] c-functions (confirmation functions) in the philosophy of probability are theoretical instruments introduced principally by Rudolf Carnap to quantify the degree to which evidence (E) supports or confirms a hypothesis (H).

observed and yet diverge in their projection. One may favour the detected uniformity more aggressively and assign high confirmation to 'the next will be F'. Another may be more conservative and keep confirmation moderate, preserving room for variation. Both may be admissible. What separates them is not a formal error. It is an inductive policy.

Here the search for impersonal confirmation reaches the problem of initial probabilities in another form. Choosing a c-function is not an act of logical discovery. It is an act of inductive legislation. It explicitly encodes how initial ignorance will be distributed and how rapidly evidence will be authorised to support generalisations. In the terms of this book, the c-function embeds an 'initial probability' of support. It is the P_0 of confirmation, without which there is no dynamics.

This observation has two effects. The first is virtuous: inductive policy ceases to operate in hiding and becomes an object of public criticism. The second is corrosive to the maximalist ambition: the policy is not fixed by pure logic. It requires additional criteria, such as invariance under renaming, stability under refinements of description and performance under idealised regimes of learning. These criteria are no longer 'pure logic'. They are methodological in nature.

At this point the problem of language appears as a problem of representation in the strong sense. Confirmation is not a brute property of the world. It is a relation defined in a language. Changing vocabulary, categories and modes of individuation can change the degree of confirmation. Logicism does not promise confirmation without form. It promises confirmation under a declared form.

The 'new riddle of induction' makes this unavoidable.

Nelson Goodman (1906–1998) shows that the choice of predicates is not an aesthetic detail. It determines what counts as a projectible generalisation. Predicates may be formally symmetrical and yet not be epistemically equivalent. The problem is not a decorative paradox. It is a structural diagnosis. Confirmation presupposes projectibility. And projectibility cannot be derived from logical form alone. It depends on how we divide the world into categories that sustain robust projections.

This vulnerability of logicism can be stated in the vocabulary of the book. Logicism faces its own problem of initial probability. The choice of language already embeds commitments about which generalisations 'deserve' to be projected. These commitments cannot be fully justified within the formal system. They come from outside it: from practices, tacit ontologies and background theories.

This opens an instructive contrast with approaches that attempt to anchor confirmation in the structure of the world rather than solely in the structure of language.

Judea Pearl (1936-) argues that many relevant inferential judgements, particularly those concerning intervention and counterfactuals, depend on causal structure. It is not enough to manipulate logical-linguistic relations among observational propositions. One must state how the world generates data. The tension with logicism is direct. Logicism seeks a normative tribunal over arguments. The causal approach suggests that part of the tribunal lies in the world, in the direction of dependencies, and that an adequate language must reflect that structure.

There is also a practical friction that logicism, by design, tends to underestimate. Carnap works with formalised languages. Actual science operates in natural language extended by formalisms. This mixture is unavoidable. Categories, hypotheses, criteria of data quality and auxiliary assumptions circulate in hybrid registers. This weakens the promise that a theory of confirmation can be applied directly as a universal calculus. But it strengthens the value of logicism as a discipline of transparency. If we cannot formalise everything, we can still declare what has been formalised and where natural language entered as a decision.

6.5 Limits of the analytic-synthetic distinction

Willard Van Orman Quine (1908–2000) is the axial critic because he challenges the terrain on which logicism relies most heavily. The Carnapian programme depends on a distinction between the analytic and the synthetic: between what would be true by meaning and what would be true by fact. If this boundary is not sharp, linguistic and theoretical choices enter the heart of confirmation, and the idea of a purely logical metric loses its frame.

Quine's criticism can be stated through an explicit mechanism. Quine argues that meaning and fact do not separate into sealed domains. Concepts acquire meaning through their role in a network of beliefs that confronts experience as a whole. If there is no linguistic core wholly immune to empirical revision, confirmation cannot be a calculation over a fixed base. It becomes a process of global readjustment. Considerations such as simplicity, cost of revision and conservatism enter as components of inferential rationality rather than as external noise.

Here it is useful to recall an antecedent that had already weakened the idea of local testing.

Pierre Duhem (1861–1916) insists that hypotheses rarely confront experience in isolation. Empirical testing bears on a whole set, including auxiliary assumptions about instrumentation, initial conditions and idealisations. Logicism, which would like a clean relation between evidence and hypothesis, encounters a structural resistance here. Support is not local. It is mediated by background theory.

With this, the ideal of a unique confirmation measure becomes difficult to sustain without remainder. Even if one fixes a formal language and a c-function, one must still decide how to deal with auxiliary assumptions, revisions of vocabulary and changes in the very theoretical framework in which hypotheses are formulated.

This kind of difficulty is also visible in paradoxes internal to the confirmation project.

Carl Hempel (1905–1997) makes explicit the tensions in attempts to formalise confirmation. The raven paradox shows how apparently plausible rules of confirmation can generate counter-intuitive consequences. What is learnt from this is not merely a technical point. Relevance, informational content and context enter precisely where logicism would like to operate with pure form.

The result need not be a surrender to arbitrariness. To say that confirmation is not purely logical is not to say that it is capricious. It is to say that inferential normativity depends on theoretical and descriptive commitments that cannot be eliminated by formal decree. The impersonality that remains possible becomes impersonality through transparency and rules of revision, not through the isolation of a self-sufficient logical core.

6.6 Cases and practical limits

In science, logicism is most convincing when it operates as a discipline of transparency and less so when it operates as metaphysics. It is especially useful when a community needs to discuss what counts as relevant evidence, how hypotheses should be formulated so that they can be confirmed, and how to prevent rhetorical preferences from replacing inferential criteria.

The vocabulary of confirmation helps to separate questions that are easily conflated. A hypothesis may fit a data set well and yet be weakly confirmed once one considers its dependence on fragile auxiliary assumptions, the use of ad hoc adjustments or the collapse of predictive power outside the observed set. In such settings, to speak of confirmation is not merely to speak of fit. It is to speak of the structure of support, inferential cost and dependence on commitments.

The notion of total evidence enters here as a regulative ideal. It does not describe how real agents always reason. It describes how they ought to reason if they seek responsible intersubjective control. The point is both normative and social. Total evidence is not merely an individual virtue. It can be institutionalised as a communal norm. Preregistration of hypotheses, disclosure of data and analytical routines, and reporting all decisions of delimitation can all be understood as attempts to realise, in practice, a logicist ideal of publicity. The community attempts to make the path intelligible, not merely the result.

Within this framework, contemporary pathologies acquire a philosophical name. Silent selection of favourable evidence, exploration of multiple analyses until a convenient result is obtained, and selection of variables after inspecting the data are all violations of the ideal of total evidence. They are degradations of publicity through selective delimitation. Logicism does not solve this problem as a calculus. It diagnoses it as a failure of inferential constitution.

The practical limit appears when we attempt to turn the whole of scientific life into a single universal calculus. Total evidence is too demanding for actual practice because information is always dispersed and there are always noise and gaps. Dependence on language weakens the promise of universality because relevant categories vary by domain. And the plurality of inductive policies suggests that a single yardstick may simply not exist.

Even so, logicism succeeds as a discipline. It compels us to declare the language, the hypotheses, the criteria and the delimitations. It turns into a public object what normally remains implicit. Even when it fails as an ultimate foundation, it supplies an ethic of disclosure that is a minimum condition for any inferential publicity.

6.7 Formal developments and their limits

The later history of confirmation shows a migration. Part of the logicist spirit is absorbed by formalisms that abandon the claim to 'unique logical probabilities' but preserve the focus on the structure of support.

One route is confirmation within a Bayesian framework, where evidence changes beliefs by an updating rule. Bayesianism can incorporate the requirement of total evidence, but it rejects the idea of an objective confirmation independent of a starting point. It transforms the problem into another one: given a declared starting point, which update is coherent?

Another route is the likelihood tradition, which seeks to measure evidential strength without requiring initial probabilities over hypotheses. Here there is an attempt to capture something like 'weight of evidence' while avoiding strong initial commitments. The gain is a reduction in the initial burden. The cost is that likelihood alone does not provide a complete policy of belief and decision.

There are also specific problems that expose the limits of any formal theory of confirmation. The problem of 'old evidence' is exemplary.

Clark Glymour (1942-) points out that evidence already known can, in some contexts, confirm new theories, which places pressure on models of confirmation that gain support only when something temporally new is learnt. The philosophical point is that confirmation is not merely a relation between data and hypothesis. It is a relation among data, hypothesis and explanatory structure. This reopens the question of how to represent 'what counts as novelty' and 'what counts as explanation'. The problem is representational once again.

Finally, attempts to measure confirmation by formal indices, whether Carnapian in style or based on alternative measures, tend to encounter the same limit. Measures compete, each with its implicit policy. The problem is not a lack of technique. It is the inevitability of a normative starting point.

6.8 Synthesis

Logicism is the interpretation that attempts to make probability a property of inferential relations. It is a measure of the support that evidence confers on a hypothesis within a language. Logicism does not begin with events, series or dispositions. It begins with arguments and normative rules.

Its strength lies in a set of mutually reinforcing moves. It separates confirmation from frequency and objective chance. It makes the role of language explicit and insists on total evidence as an ethic of inferential control. It makes visible what is often concealed in actual scientific practice. There is always a policy of description. There is always a policy of projection. Publicity does not arise from silence about these points. It arises from exposing them.

The price of this virtue is equally clear. The programme depends on choices about language and inductive policies that are not fixed by pure logic. Goodman's lesson suggests that projectibility precedes confirmation. Quine's criticism suggests that meaning and fact are interwoven in a network, dissolving the dream of an immune analytic core. When these criticisms are taken seriously, the idea of a single impersonal confirmation loses plausibility. Impersonality shifts. It ceases to be 'one correct number' prior to the subject and becomes a regime of transparency, revision and communal discipline.

This shift prepares the next chapter through a dialectical movement. Logicism sought a normative tribunal external to the subject, a public calculus of support. But it cannot eliminate the starting point. It merely reconfigures it as a choice of language and inductive policy. In response, a radical alternative gains force: abandon the search for 'objective' probabilities in the logical sense and internalise normativity. If there is no external logical tribunal that fixes the correct confirmation, rationality may reside in the internal coherence of the belief system and in the discipline of updating.

Here the historical bridge must be stated through names and frictions.

Frank Plumpton Ramsey (1903–1930) serves as the link because he accepts the demand for rationality but anchors it in consistency of preference and decision rather than in a Platonic inductive logic. Bruno de Finetti (1906–1985) explicitly breaks with the idea of objective logical probability. For him, probability is degree of belief, and its discipline is coherence. The tension with Carnap is structural. Carnap sought a public metric of confirmation through language. De Finetti insists that this 'public' value does not exist as a fixed number, only as agreement and as a consistency constraint.

The decision-theoretic architecture that makes this programme operational takes shape with Leonard Jimmie Savage (1917–1971), who binds beliefs and preferences under axioms. The core of subjectivism, then, will not be 'probability as impersonal logical support'. It will be 'probability as coherent degree of belief'. The promise is attractive. If there is no external tribunal capable of fixing a unique confirmation, there can at least be an internal discipline that prevents incoherence.

Objectivity will not be given as an impersonal number. It will have to be reconstructed afterwards, through convergence under shared evidence, through communal norms of transparency, or through principles linking degree of belief and objective chance. In every case, publicity will have to be recovered as a construction. Never as a given.

The next chapter begins precisely here. Logic ceases to promise a public number that precedes the agent. Normativity takes up residence within the agent. And the initial probability, formerly an embarrassing constraint, becomes the declared point at which inferential ethics begins.

7 Subjective Bayesian Interpretation

(From impersonal confirmation to the discipline of belief)

After logicism, probability can no longer pretend to exist outside the agent. John Maynard Keynes (1883–1946) and Rudolf Carnap (1891–1970) sought an impersonal confirmation tied to logical form and formal language. The project taught us much, but it left a difficult remainder. The objectivity it sought requires prior choices: choices of language, choices about what counts as total evidence, choices of inductive policies. Willard Van Orman Quine's (1908–2000) criticism merely made unavoidable what was already in the air. There is no living inference without a place at which epistemic decisions are made.

Bayesian subjectivism is not a concession to epistemic relativism. It is a radical change in the location of normativity. Instead of requiring, as a starting point, a direct fit between belief and some privileged external fact, whether truth, frequency or causal disposition, it shifts the primary criterion to the integrity of the architecture of belief itself: coherence, discipline in updating and conditions of public control. The question changes. Instead of asking 'what is the true probability?', one asks 'what makes a probabilistic belief rational, stable and publicly controllable?'. When objectivity appears, it appears as the product of an integral and traceable process, not as a brute datum prior to the process.

The change is internal, but it does not seal the world inside the agent. The world returns as an indirect tribunal: through evidence, predictive failures, model revisions and, where appropriate, through the co-ordination of degree of belief with objective chance. The difference is that this tribunal does not operate as immediate access to a sacred number. It operates as a progressive constraint on a declared inferential practice.

7.1 Bayesianism in the intellectual climate of the twentieth century

The resurgence of Bayesianism in the twentieth century did not occur in a vacuum. It developed in an environment in which the limits of wholly impersonal programmes of inference and the underdetermination of theories by data became increasingly visible. In philosophy of science, the Duhem-Quine problem emphasises that hypotheses are not tested in isolation and that auxiliary assumptions participate in inference. This background did not, by itself, cause the return of Bayesianism, but it helps to explain why making assumptions, initial probabilities and updating rules explicit came to be seen as a methodological virtue rather than merely a subjective concession.

Against this background, the Bayesian gain is both methodological and ethical. What cannot be eliminated should be exposed. What cannot be made impersonal by decree should be made public through traceability. The initial probability becomes a declaration and, for that very reason, a legitimate target of criticism.

7.2 Degree of belief and coherence

Frank Plumpton Ramsey (1903–1930) introduces a connection that might sound reductive but is constitutive. If someone claims to believe by degrees, those degrees must have content that organises choices under risk. A disposition to accept certain bets is not an external sign of belief, as though it were a psychological clue. For Ramsey, it is part of the very meaning of 'having a degree of belief'. To say 'I believe H more than K' cannot remain internal poetry. It must manifest itself in how one prefers to act when something is at stake.

A minimal distinction should be fixed in order to avoid loose psychologising. A degree of belief is a graded inferential commitment, normally represented by numbers between 0 and 1 assigned to hypotheses under partial information. It is not an emotion. It is a practical rule of judgement capable of guiding prediction, decision and revision.

Coherence enters as a condition of possibility. It is not prudential advice of the form 'be careful'. It is a constitutive condition. A system of probabilistic beliefs deserves that name only if it does not permit the construction of a set of bets against it that guarantees a loss whatever happens. When this is possible, the loss is certain by construction. It is not bad luck. It is logical-practical incoherence. The architecture of belief has short-circuited.

Here it is worth separating two layers that are often fused. Coherence disciplines the synchronic state of degrees of belief. It is a static requirement. Dynamic discipline—how to change a degree of belief in the light of evidence—is an additional step. Bayesianism claims a norm of revision that avoids both dogmatic rigidity and opportunistic plasticity.

Without this, coherence becomes merely internal hygiene rather than a method of learning.

This distinction touches a deep problem. The Bayesian norm is strong precisely because it makes explicit that learning is not a miracle. It is a declared rule. But this strength has a cost. The rule requires us to state what the evidence was, what the model was and what the initial probabilities were. The price of discipline is exposure.

In addition to Dutch Book arguments, an independent line associated with James Joyce and developed by Richard Pettigrew justifies probabilism through considerations of epistemic accuracy. On this approach, proper scoring rules evaluate credences according to their closeness to truth and support dominance results: systems that violate probabilistic norms can be outperformed by more accurate alternatives. The philosophical importance of this strategy lies in showing that the defence of probabilism does not depend exclusively on a pragmatic interpretation in terms of betting (Joyce, 1998; Pettigrew, 2016).

7.3 Exchangeability and structural objectivity

Bruno de Finetti (1906–1985) radicalises the thesis. His dictum, 'probability does not exist', denies neither calculation nor usefulness. It denies an automatic ontology. Probability is not an entity inhabiting events like an invisible weight. It is a disciplined way of expressing uncertainty and revising it without contradiction.

Coherence becomes the primary tribunal. A set of probabilistic assessments is acceptable if it does not allow guaranteed defeat by construction. This is a normative and public criterion, not a psychological diagnosis. It defines the minimum required for an agent to call their own judgement rational.

Then comes the concept that brings subjectivism into contact with a promise of practical objectivity. Exchangeability is subtler than independence and does not imply it. It captures an epistemic symmetry. The order of outcomes is judged irrelevant for purposes of learning. What matters is the composition of what has been observed and what it permits one to infer. This is a decision about informational irrelevance, not a metaphysical claim about causal mechanisms 'without memory'.

With this commitment comes a representation result. In its classical form for infinite sequences of Bernoulli variables, de Finetti's theorem states, in broad terms, that an exchangeable sequence can be represented as a mixture of conditionally independent and identically distributed sequences. The mixing measure acts as a distribution over a long-run latent parameter. The philosophical reading should be sober: the representation does not prove that this parameter exists as a physical entity; it shows that certain exchangeable assessments admit a mathematical form equivalent to a mixture of i.i.d. models (de Finetti, 1974-1975; Aldous, 1985).

Representation must also be distinguished from convergence between agents. De Finetti's theorem alone does not guarantee that two agents with different initial probabilities will eventually agree. Merging results, such as that of Blackwell and Dubins, establish approximation of predictive distributions under additional conditions, including appropriate absolute-continuity relations between the initial measures. 'Objectivity through convergence' is therefore a conditional possibility, not an automatic consequence of exchangeability (Blackwell; Dubins, 1962).

7.4 Initial probabilities and updating

Subjectivism gains power at the cost of exposing its most uncomfortable component. The beginning matters. The agent enters the scene with prior beliefs. This starting point is expressed by initial probabilities. It is unavoidable, even when one attempts to hide it. When someone says that they have no initial probabilities, they have almost always merely placed them silently inside the model, in choices of parametrisation or in what they have called 'neutral'.

Here it is useful to recover the historical thread on a short, functional scale. Thomas Bayes (1701–1761) opens a formal route for updating degrees of belief by evidence, and Pierre-Simon Laplace (1749–1827) expands and popularises the practice of inverse probability in scientific problems. At its origin, however, the gesture still carries an ambition to impersonality through the principle of indifference. The nineteenth century, by criticising that principle and consolidating frequentist readings, pushes inverse probability into a relative eclipse. The twentieth century returns to the subject with a different grammar. It does not promise a beginning without choices. It promises a declared and disciplined beginning.

The updating rule is the core. It is not a matter of creating belief from nothing upon seeing data. It is a matter of redistributing belief in the light of evidence, moving towards hypotheses that make the evidence less surprising and away from hypotheses that made it improbable. The method does not promise immunity from error. It promises that error will not be protected by incoherence or methodological silence.

The dilemma of the beginning and the illusion of a zero point

The friction is direct. Different initial probabilities can produce different conclusions from the same evidence, especially in small samples, complex models or when the evidence is weak. The problem is not merely practical. It is philosophical. If probability is degree of belief, why should the initial degree not be merely a matter of epistemic taste?

The honest answer begins by dismantling the illusion of a zero point. There is no inference without assumptions. The task is not to eliminate assumptions. It is to make the relevant assumptions explicit, assess them, test them through sensitivity analysis and submit them to public control. In this book, subjectivism replaces the fantasy of absolute neutrality with explicit responsibility. It exchanges a hidden beginning for a criticisable one.

Convergence, calibration and co-ordination with chance

Subjectivism does not defend itself by promising instant unanimity. It defends itself through earned objectivity. Part of that objectivity appears as convergence under evidence. In many contexts, reasonable initial probabilities tend to lose influence as the evidence grows. This is not magic. It requires openness to learning. It also requires that the beginning not freeze genuine possibilities out of consideration.

Another part appears as predictive calibration. Even where there is no ontological agreement, there can be public evaluation through performance. An inferential practice is better when it produces predictions that hold up, when it fails in diagnosable ways, and when it exposes what depended on the beginning and what was determined by the data.

When there are reasons to speak of objective chance, whether through stabilised frequency or causal mechanism, rational degree of belief should not ignore it. Under appropriate conditions it should be co-ordinated with that chance, as David Lewis

(1941–2001) formulates. The point in this book is not to declare a metaphysical winner. It is to prevent unilateral sovereignty. Chance without degree of belief does not decide; degree of belief without constraint from the world becomes methodological narcissism. Bayesian discipline is one way of keeping this tension under public rules.

7.5 Internal objectivity and criticisms

'Subjectivism' is a broad label. Within it there is a spectrum that is not a sign of weakness but of philosophical maturity. The term covers a genuine tension between poles.

At one end there is a pole of epistemic freedom, where coherence is the minimum and almost sufficient norm. Objectivity emerges as intersubjectivity constructed through common practices, common data and explicit criticism.

At the other end there is a pole of constraint internal to the formalism, where one attempts to restrict initial probabilities through formal principles, invariances and criteria of 'non-informativeness'. Harold Jeffreys (1891–1989) and Edwin Thompson Jaynes (1922–1998) belong here in different ways. This is a reenactment, in the domain of beginnings, of the old desire for a natural measure. The ghost of Joseph Bertrand (1822–1900) returns in new clothing. The problem of space and metric does not disappear. It changes location and continues to require choices about parametrisation, relevant symmetry and reference measure.

Between these extremes lies the pole of regulated practice. Objectivity is not sought in a unique beginning, but in protocols of transparency, sensitivity analysis, predictive calibration, communal criticism and replicable performance. Objectivity becomes procedural integrity. Absolute neutrality is not promised. What is promised is control of inferential risk through exposure and revision.

Realist criticisms attack the programme along three lines. It is said that coherence does not guarantee truth. It is said that there are domains in which objective chances seem indispensable. It is said that dependence on the beginning is a vice. The Bayesian response separates levels. Coherence is not a guarantee of truth. It is a condition ensuring that error is not protected by internal contradiction. Objective chances may exist and yet the agent's problem remains: how to co-ordinate degree of belief with partial evidence about those chances. And the beginning is not a defect. It is the visible place where assumptions enter and where they can be criticised without pretence.

7.6 Persistent philosophical challenges

Some challenges are not peripheral noise. They are structural.

The first is the problem of the hypothesis space. What counts as an admissible hypothesis defines what can be learnt. If the hypothesis space already excludes genuine alternatives, updating merely refines an initial mistake. Here Bertrand's problem returns in abstract form. The issue is not merely how to choose a beginning within a given space. It is how to justify the space itself.

There is also a challenge to the very demand for point precision. Theories of imprecise probabilities allow states of information to be represented by intervals, lower and upper probabilities, or sets of distributions, rather than forcing a single numerical credence where the evidence does not support one. These approaches neither abolish Bayesianism nor constitute, by themselves, a single interpretation of probability; they show that the choice of the format of uncertainty also has epistemological content (Levi, 1980; Walley, 1991).

The second is the problem of old evidence. There are situations in which evidence already known comes to play a confirmatory role for a new theory. This is a severe test for the simple narrative of updating because the evidence does not arrive as psychological novelty. Its function changes when the theoretical language changes.

The third is the problem of complexity. In high-dimensional spaces, the beginning can exert effects that are difficult to diagnose. The risk is to exchange the illusion of a zero point for an invisible engineering of the starting point. For this reason, the criterion in this book is not coherence alone. It is auditability, sensitivity and predictive performance under scrutiny.

The fourth is the status of scientific realism. Bayesianism organises degrees of belief. By itself, it does not guarantee that our best theories capture the world rather than merely regularities in what has been observed. This is not a defect in the calculus. It is the tension between predictive success and ontological commitment, now reenacted in probabilistic form.

7.7 Decision and scientific practice

In practical decision-making under uncertainty, subjectivism reveals its strength through epistemic honesty. A doctor estimates risk, weighs harms and chooses a course of action. The patient is singular. The question is singular. Probability in this context cannot be merely the frequency of a distant population without an explicit reference class and an explicit translation from the singular case to the series. It must represent the best available degree of belief, conditional on evidence, models and experience. The virtue here is not omnipotence. It is traceability.

In scientific inference, the method organises the confrontation among hypotheses, models and data. The scientist does not extract facts as though reading them directly. They construct explanations, update plausibilities, identify what the data genuinely inform and what the model imposes. The cultural gain is that uncertainty enters the report instead of remaining a hidden residue.

The ethical dimension appears when we remember that beliefs have costs and evidence is finite. The classic debate between William Kingdon Clifford (1845–1879) and William James (1842–1910) reappears here in another grammar. Bayesianism does not resolve the conflict by moral decree. It transforms it into explicit responsibility. If you choose a beginning, you declare it. If you update, you show the path. The obligation is no longer to pretend to impersonality. It is not to conceal assumptions.

7.8 The battle of the schools and the computational revolution

In the twentieth century, applied statistics was shaped to a large extent by a frequentist hegemony associated with Ronald Aylmer Fisher (1890–1962), Jerzy Neyman (1894–1981) and Egon Sharpe Pearson (1895–1980). This was not merely a philosophical dispute. It was an institutional choice of language, reporting criteria and protocols. For long periods, Bayesianism remained a minority tradition, sustained by arguments from coherence and by a unified view of inference and decision.

The modern turn has two engines. One is conceptual, linking inference to decision and giving explicit centrality to the beginning. The other is practical, making computable on a large scale what had previously been elegant but infeasible. The spread of Markov chain Monte Carlo methods in scientific practice changes the landscape. Bayesianism becomes an everyday tool in complex models, and the publicity of calculation comes to depend also on code, reproducibility and computational audit.

This closes a circle with the central thesis of this book. In the twenty-first century, objectivity is increasingly less a property of an isolated number and increasingly more a property of a replicable path.

7.9 Synthesis

Bayesian subjectivism answers two earlier impasses.

To logicism, it responds by accepting that confirmation cannot be obtained as a unique relation between propositions independent of the agent and their conceptual apparatus. The objectivity dreamt of in pure language cannot be realised without convention or cost. Bayesianism exchanges the dream of a unique measure for a requirement of dynamic coherence.

To propensity theory, it responds by recognising that, even if the world has tendencies expressible as objective chances, a finite agent must learn under partial evidence. The co-ordination between chance and degree of belief is not automatic. It requires rules, updating and discipline. This is where Bayesianism finds its necessity.

Bayesian subjectivism therefore offers a broad and internally disciplined interpretation of probability as degree of belief. Its conceptual cost is to accept the inevitability and visibility of the starting point. In return, it provides a unified structure for dynamic rationality under uncertainty, capable of incorporating evidence of different kinds and of making every step of the reasoning explicit.

This victory of the normativity of coherence over an automatic metaphysics of probability does not end the debate. It raises it to another level. The question becomes how to co-ordinate, without unilateral sovereignty, different ways of speaking about probability—frequentist, propensity-based, Bayesian—within a pluralist and responsible scientific practice. It is in this arena of pluralism regulated by practice, where interpretations cease to be rival doctrines and become tools under collective constraints of inquiry, that the conceptual journey of this book reaches its culmination.

8 Criterion-Guided Pluralism

(How to manage trade-offs without relativism)

After traversing the major interpretations, it is tempting to treat them as political parties, each with its banner, its heroes and its scandals. That framing falsifies the intellectual history reconstructed here. The real conflict is not among 'schools' as matters of taste. It is among constitutive requirements of rationality under uncertainty that prove, in practice, to be structurally in tension. We want probability to be objective and useful in the singular case. We also want it to be inferentially fruitful and publicly auditable. We want it to be stable when we change the description and sensitive to what the evidence imposes. We want it to serve explanation and the rational calculation of decision. The methodological conflict begins when someone promises to maximise all these virtues at once, as though they were naturally compatible.

This chapter argues that criterion-guided pluralism is the most adequate response to the tensions reconstructed thus far. The conclusion follows not merely from the existence of different schools, but from the recognition that the principal interpretations distribute advantages such as publicity, applicability to singular cases, explanatory force and inferential discipline in different ways. This does not prove that every monism is logically impossible; it does, however, require any proposal for uniqueness to show how it manages these costs without concealing them. The pluralism here is semantic and methodological. We do not pluralise the standard calculus. We pluralise the regimes of justification and interpretation that connect the formalism to the world, evidence and decision.

8.1 Tensions and methodological conflict

The interpretive conflict unfolds along irreducible axes of tension. These are not topics that can be resolved through better rhetoric. They are constitutive dilemmas of probabilistic rationality. With every attempt at foundation, they reappear, because they express what we simultaneously want from probability and what the world and inferential practice rarely allow us to obtain all at once.

The first axis concerns where objectivity should reside. Frequentism locates objectivity in public long-run performance, in auditable guarantees and in technologies of error control. Propensity theory seeks objectivity in the world, as objective chance or a disposition of a causal set-up. Logicism seeks it in the impersonal validity of inferential relations. Subjectivism seeks objectivity through the discipline of the agent, relying on transparency, calibration and convergence. The costs appear immediately. Operational objectivity is publicly measurable but tends to mediate the singular case through classes and procedures. Metaphysical objectivity promises to say what exists in the world but pays through ontological commitments and the problem of co-ordinating objective chance with degree of belief. Normative objectivity has conceptual elegance but suffers from dependence on language, conventions and intractability. Objectivity as discipline of the subject gains inferential power in the singular case but exacts responsibility for the starting point. Here, 'objectivity' is a family-resemblance term. It designates different standards of justification—auditable publicity, ontological commitment, normative impersonality, disciplined coherence—that cannot be reduced to one another without loss. The common feature is always some form of publicity: of the world (chances), of the method (performance), of the language (rules), or of the inferential path (declaration and auditability). This plurality of standards aligns with epistemic pluralism but is specified here within the probabilistic domain. It is not merely a matter of accepting multiple valid descriptions. It is a matter of recognising that different inferential contexts legitimate different forms of public justification.

The second axis concerns the place of the singular case. Frequentism has a clear public semantics, but that clarity has a price. Frequentist objectivity is, by definition, a property of the population and the method, not of the unique event precisely as unique. Subjectivism and propensity theory apply directly to the singular case, but the application incurs costs. In Bayesianism, the costs are explicit initial probabilities and the integrity of the inferential process. In propensity theory, the additional cost is the step that turns objective dispositions into legitimate inferences without dissolving causal asymmetry into the symmetry of the calculus. The trade-off here is not between 'right' and 'wrong'. It is between robust public semantics and direct applicability to the single case.

The third axis concerns the status of evidence. The trade-off appears when we operationalise 'evidence' as though it were a single object. In fact, it is a multifaceted concept. It may be raw data. It may be the result of a standardised procedure. It may be an inferred causal trace. It may be a formal constraint on belief updating. Each interpretation tends to select and prioritise one of these facets as primary, and conflict erupts when the facet privileged by one programme is treated as though it exhausted what 'evidence' means in scientific practice.

The fourth axis concerns what probabilistic calculation is doing. On some readings, it functions as a technology for updating degrees of belief and deciding under risk. On others, it forms part of models intended to represent causal relations or logical-confirmational relations. The distinction is decisive: probabilistic calculus by itself contains no causal arrow; causal structures require additional assumptions about interventions, mechanisms or independences. The same formalism can serve different metaphysical and normative projects, and confusion begins when the semantics of one of those projects is imported without declaring the corresponding bridge.

Two test cases make this diagnosis less programmatic and more verifiable. In a randomised clinical trial with a binary outcome and a replicable protocol, the central advantage is the publicity of the procedure: error control, interval coverage, performance under repetitions of the same design. The frequentist reading fits naturally, and the demand for 'probability in the singular case' is met indirectly, as a property of the method applied to a defined population. By contrast, in the individual risk of a rare patient, with sparse data and a specific clinical context, the essential requirement is finite-sample inference. The probabilistic number must incorporate prior information, plausibility of mechanisms and sensitivity to model uncertainty. Here the central virtue shifts to transparency about initial probabilities, calibration and sensitivity analysis, under an obligation to make the chain of decisions public.

The history of the interpretations traversed in Chapters 3 to 7 is therefore not a sequence of errors to be overcome, but a map of attempts to resolve these tensions. Each interpretation chose a point in this space of trade-offs at which to anchor itself, maximising certain advantages while accepting the corresponding cost. Conflict arises when one forgets that the anchor is a choice and often a methodological choice, not an inevitable discovery. Faced with these constitutive tensions, strategies of resolution emerge that seek to subordinate one dimension to another. These are the projects of uniqueness examined next.

8.2 Projects of uniqueness

There is a clear strategic kinship among Hans Reichenbach (1891–1953), Karl Popper (1902–1994) and Bruno de Finetti (1906–1985). The move is the same in the face of trade-offs. One axis of response—methodological, realist or normative—is isolated and elevated to the status of exclusive foundation, while the remaining demands are relegated to derived problems to be managed ad hoc. The kinship lies not in content but in the form of the strategy.

In Hans Reichenbach (1891–1953), uniqueness is methodological. We act in the finite case as though there were a stable limit behind the sample, not through metaphysical contemplation but because, without that leap, there is no workable induction. The posit attempts to tame finitude. The cost reappears in the choice of reference class. Operability is gained and neutrality in delimitation is lost, because the 'objective' number exists only after a criterion for the relevant series has been chosen.

In Karl Popper (1902–1994), uniqueness is realist. Propensity theory promises to save the singular case by locating objective chance in the experimental set-up. Here the cost takes another form. The ontology of dispositions is asymmetric because causes and effects are not mirror images. The probability calculus, as algebra, is symmetrical in the inferential sense. The propensity project requires a principle of co-ordination to prevent formal symmetry from being confused with causal content. Without such co-ordination, metaphysical uniqueness does not translate into public inference.

In Bruno de Finetti (1906–1985), uniqueness is normative. Probability does not exist as an entity in the world, but there is an inescapable requirement of coherence for any agent who wishes to speak meaningfully in degrees of belief. Uniqueness here is not a 'true' number. It is a standard of consistency that prevents contradictions and guaranteed losses. The cost is clear. Coherence is necessary, but it is not an automatic link with the world. The bridge must come from learning, predictive performance, sensitivity analysis and public practices of audit.

These searches for uniqueness are not redundant. Each attempts to master one central trade-off from Section 8.1 by shifting the cost to another axis. This is why criterion-guided pluralism does not arise from tolerance. It arises from the diagnosis that total uniqueness is always purchased at a price and that the philosophical duty is to make that price visible. Chapter 9 will show how these costs reappear, not as disputes among schools, but as typical failures in concrete fields.

8.3 Public constraints

Objective Bayesianism is an attempt at reconciliation. It accepts the discipline of updating and the rationality of the agent but seeks to constrain initial probabilities through publicly defensible principles: invariance, symmetry, maximum entropy under constraints, reference distributions. The impulse is understandable. It seeks to reduce the discomfort of the problem of initial probabilities without abandoning the Bayesian machinery.

The difficulty appears when we ask where the decision is hidden. The paradigmatic case is maximum entropy. The promise is simple. Given a family of distributions compatible with certain constraints, one chooses the distribution that 'assumes no more than necessary'. Reality is more austere. 'Not assuming' depends on the hypothesis space and on a base measure that defines what counts as 'spreading uniformly'. If I describe the same problem by a parameter or by a transformation of it, what appears neutral in one system ceases to appear neutral in the other. Choosing the base measure, choosing the relevant constraints, choosing the delimitation of the hypothesis space itself: each step decides the result.

At this point, the logical structure of Joseph Bertrand's (1822–1900) paradox reappears in new clothing. Choosing a 'natural measure' for maximising entropy is no more neutral or unique than choosing a method for 'selecting a chord at random'. Far from being resolved, the problem of measure and parametrisation is re-internalised at the heart of the formalism described as 'objective'.

This does not make objective Bayesianism useless. In domains with justifiable symmetries and representations stabilised by theory and practice, it can be excellent. Its advantage is to make the discussion of constraints more public and articulable. Its risk is to mask metatheoretical decisions beneath an aura of technical neutrality. Objective Bayesianism therefore does not solve the problem of initial probabilities. It shifts it to a metatheoretical level, at which choices of invariance, reference measures and notions of 'minimal information' must be defended as responsible decisions rather than treated as inevitable derivations.

8.4 Hybrids and criteria for use

Hybrids are not patches. In many domains, they are the most honest form of probabilistic rationality. The scientific world combines mechanisms, evidence, models, costs, instruments and communities. The danger does not lie in combining them. It lies in combining them without declaration and in using an inferential product as though it possessed a semantics it does not have.

Criterion-guided pluralism operates as a decision tool. It is organised as a sequence of questions and nameable principles that can be cited, audited and taught. By a 'regime' we mean a co-ordinated set of semantics attributed to probability, accepted standards of justification, criteria of inferential success and protocols of public criticism. For example, the regime of a randomised clinical trial co-ordinates frequentist semantics (coverage and error control), public justification (protocol and replication), criteria of success (error targets and intervals) and criticism (peer review and reanalysis).

These criteria operate in cascade. The purpose determines the priority advantages. The epistemic regime indicates which tools formally realise them. Ontological prudence prevents category confusions. Radical transparency allows the entire path to be audited. Violation of any link compromises inferential legitimacy.

First criterion: purpose. Is the goal causal explanation, calibrated prediction, decision under risk, or some combination of these purposes? The answer already defines which advantages should be maximised and which costs will be acceptable. Calibrated prediction requires predictive testing and stability. Causal explanation requires attention to mechanisms and asymmetries. Decision under risk requires clarity about losses and utilities.

Second criterion: the epistemic regime. Does the problem live under controlled repetition and public audit, or under irreducible singularity and sparse data? Where controlled repetition is available, frequentist justification is especially strong because it provides public guarantees about the reliability of the method. Where it is not, inference must explicitly assume finitude and singularity. In that case, the Bayesian tool is often more natural, provided that transparency, calibration and sensitivity analysis are treated as obligations.

Third criterion: ontological prudence. Is there robust plausibility of objective chances relevant to explanation, or are we dealing predominantly with epistemic uncertainty about the model, omitted variables and mechanisms? In certain physical domains, speaking of objective chance can organise theory powerfully. In many complex domains, calling 'chance' what is, in practice, epistemic uncertainty tends to create confusion. Conceptual prudence here is part of scientific honesty.

Fourth criterion: radical transparency. Whatever the preceding choices, structural decisions must be declared: reference class, model, auxiliary assumptions, initial probabilities, base measure, invariance criteria and parametrisation. Auditability of the path is treated here as a public minimum of methodological objectivity, not as an exhaustive definition of objectivity. A transparent procedure can still be misspecified or begin from inadequate categories; transparency must therefore be accompanied by the other criteria of purpose, regime and ontological prudence. The minimum point is that a probabilistic assignment that cannot be reconstructed and criticised by competent agents does not satisfy even the basic public condition of inferential legitimacy.

These criteria are not a checklist. They formalise requirements of situated rationality. They adjust inferential standards to the purpose of the problem, the epistemic resources available, defensible ontological commitments and conditions of public criticism. Their normative character does not derive from a priori deduction but from their capacity to reduce foreseeable epistemic failures, such as semantic slippage, confusion between epistemic uncertainty and ontological chance, and the improper application of long-run guarantees to unique, unrepeatable cases.

We call the requirement of semantic consistency the inferential golden rule. The justification supplied for a probabilistic number must govern its interpretation and the scope of its conclusions. It is illegitimate to take the justification from one interpretive regime—for example, a frequentist guarantee of coverage—and take the conclusion from another—for example, a degree of belief about a singular parameter—without an explicit and legitimate bridge. This rule is intended to eliminate the shifts of meaning that generate much of the misunderstanding and abuse in applied science. The target is simple. To calculate a 95% confidence interval as a frequentist guarantee of coverage under repetitions and then claim that, for the experiment actually performed, 'there is a 95% probability that the parameter lies in this interval' is an illegitimate transposition of semantics. By contrast, saying 'under this Bayesian model, with these initial probabilities and these assumptions, my posterior credence is 90% that the parameter lies in this interval' satisfies the requirement, because the justification—updating under a declared model and starting point—governs the interpretation—degree of belief. Likewise, calculating a 'probability of causation' from a purely associational model violates this requirement by assigning causal semantics to a relation that the model, by construction, cannot sustain.

At this point it is useful to name what has already been described. The set of requirements above forms an Inferential Declaration Protocol, not yet given a stable formalisation. It will be applied in Chapter 9 as a discipline against slippages of meaning and consolidated in Chapter 10 as a public standard of accountability.

8.5 Responses to objections

A predictable objection concerns incoherence: how can interpretations be combined without contradiction? The answer is simple. Interpretations are not combined as though they were semantic additives. Distinct justificatory regimes are applied to distinct components of the problem, with declared bridges when one moves from one regime to another. Frequentist simulations can audit the performance of an inferential procedure within Bayesian models, for example, without converting coverage into posterior belief.

Another recurrent objection is that 'anything goes': pluralism would lead to relativism. The answer here is structural. The requirement of semantic consistency and the demand for radical transparency are rigid constraints. This is not a matter of individual preference, but of making the inferential path public. If pluralism led to relativism, the requirement of semantic consistency itself would be optional. But that requirement is constitutive of probabilistic rationality. Without it, the same formalism would mean everything and nothing, destroying the possibility of scientific communication. What pluralism recognises is that there are multiple ways of satisfying this requirement, each with its own trade-offs.

When these principles operate, pluralism ceases to look like a catalogue of tolerance. It becomes a methodological ethic. It recognises that probabilistic rationality comprises partly incompatible benefits and that good science consists in choosing which virtues to prioritise, under which commitments, with which safeguards and through which mechanisms of public criticism.

8.6 Synthesis

The course of the book has shown a sequence of attempts to house objectivity. In the classical interpretation, it was sought in rational ignorance and returned as a problem of space and measure. In frequentism, it was anchored in the public performance of procedures and reappeared as the drama of the singular case and the reference class. In propensity theory, it was reinstalled in the world and demanded co-ordination with inference and degree of belief. In logicism, it was sought in impersonal validity and revealed a dependence on language, conventions and practical limits. In subjectivism, it re-emerged as discipline of the agent and procedural integrity, paying the price of responsibility for initial probabilities and the demand for publicity.

The cross-cutting debates crystallise the final consequence. There is no single place where objectivity can reside without cost. The mature task is not to choose a dogma. It is to make the cost explicit, choose by criteria and prevent illegitimate transpositions.

This chapter therefore offers not a truce in the conflict, but its transformation. The dispute ceases to be among interpretations treated as wholes and moves inside each complex scientific problem. Criterion-guided pluralism is a manual for this new kind of inferential management. Its decisive test, however, is still to come. It will occur not only in theory but in scientific practice. In domains where lives and knowledge are at stake—from quantum theory to epidemiological risk, from financial markets to social judgements—the interpretations will be forced to leave their abstract kingdoms. There, the choice of a reference class, the parametrisation of initial probabilities and the reading of a p-value are not academic exercises. They are constitutive acts of inference that shape the world. The decisive test will come when we confront problems such as causal inference from observational data with unmeasured confounders, where the trade-offs among frequentist guarantees, Bayesian modelling and causal interpretation become acute and inescapable.

The confidence of this pluralism, therefore, will not be a confidence of abstract correspondence. It will be a pragmatic confidence of navigability. It will be the adequacy of the conceptual map to the inferential territory, assessed by its ability to guide us effectively through real problems without epistemic shipwrecks such as contradictions, semantic slippages and systematically failed predictions. Our conceptual map will be judged not by its abstract elegance but by its capacity to guide us through the rough terrain of actual inference. It remains to be seen whether this map withstands contact with concrete cases and what kind of epistemic traveller it forms us to become. Our journey now turns to this decisive test, on the terrain of concrete cases. Probability is public only when the path is public.

9 Analytical Application of Criterion-Guided Pluralism

(Where probability decides and where semantics slips)

This chapter applies criterion-guided pluralism to domains in which probability guides substantive decisions and in which, for that very reason, semantic transpositions incur significant epistemic and practical costs. The aim is not to elect a winning interpretation, but to make explicit the constitutive epistemological tensions imposed by each field, specify which layer is active in each use, and identify typical slippages, understood here as categorical errors of semantic transposition through which a number acquires the appearance of saying more than it can say.

9.1 Operational conventions of the chapter

We shall fix a convention for the use of 'objectivity'. We take it in an operational sense, as auditability of the path, that is, as the possibility of reconstructing, criticising and repeating, under public rules, the chain of decisions that led to the number. This operational definition does not deny that, in specific substantive contexts, one may legitimately speak of 'objective chances' as properties of the world, as propensity interpretations maintain. It does, however, fix the central methodological point of this book. From the standpoint of the public justification of an inferential result, the objectivity that can be required is the objectivity of the path. Even a claim about an objective chance becomes scientifically legitimate only when the process that identifies it is itself objective in this operational sense: reconstructible, criticisable and repeatable under public rules. Thus, 'objectivity of the path' functions here as a required public minimum, that is, as a criterion of public legitimacy rather than a metaphysical thesis against chances.

The second convention is the Layer Principle, which separates three tasks. The semantic-ontological layer states what the number purports to assert about the world. The methodological-epistemic layer states how the number is obtained under finite evidence, with models and initial probabilities. The intersubjective-control layer states what is publicly audited in the procedure, especially systematic error, stability, calibration and reproducibility. These three layers correspond, respectively, to semantics (semantic-ontological), inference (methodological-epistemic) and audit (intersubjective control), in the vocabulary of Chapter 8. A minimal example fixes the distinction without prolonging the introduction. When one says that 'risk is 10%', this may be an assertion about the world, an inference under data and model, or a certification that a procedure is calibrated under a protocol. The fallacy occurs when one moves from one reading to another without justifying the move. From this point onward, whenever a passage 'appears' to say something about reality, we shall indicate whether it speaks about the world, the mode of learning under data, or the public control of the procedure.

9.2 Map of epistemic navigation

This map serves as orientation for the journey that follows. We begin with quantum physics, where the tension between ontology and inference puts acute pressure on the layers and makes rigorous separation unavoidable. Next, in statistical mechanics, the problem shifts to the legitimation of emergent descriptions, requiring vigilance over delimitation, reference measure and regime assumptions. Population genetics introduces the complexity of distinguishing process randomness, parametric uncertainty and model uncertainty, revealing the risk of converting counterfactual relations into essential attributes. In epidemiology, the tension acquires an immediate ethical cost, because transposition between population scale and singular decision tends to generate unwarranted promises. Economics and finance expose the structural epistemic cost of reflexivity, which erodes the stability of the regime itself and renders long-run guarantees fragile. Finally, the social sciences force us back to the beginning of the path, showing that criticism must start with the categorical delimitation that defines the space of possibilities and conditions what will count as evidence. Each stage puts pressure on a specific link in the inferential chain and reveals a characteristic slippage.

9.3 Quantum physics

Here the criterion of ontological prudence from Chapter 8 is placed under particular pressure because the semantic-ontological layer is contested while, at the same time, probabilistic numbers lie at the heart of experimental practice. In quantum physics, the demand for ontological prudence does not stem from an unfounded metaphysical preference, but from the way the probabilistic number appears in theory and experimental practice. The semantic-ontological layer, which states what the number says about the world, is especially strained here. The methodological-epistemic layer, which governs inference under finite evidence, appears as an unavoidable discipline in the reading and use of data. The intersubjective-control layer, which certifies procedures and apparatuses, operates as a condition for the publicity of the experiment.

Karl Popper (1902–1994), in a propensity framework, represents the attempt to anchor chances in the experimental set-up—that is, in preparation, interaction, apparatus and recording rule—thereby reducing ambiguity between physical description and the language of updating. By contrast, the Copenhagen interpretation, associated with Niels Bohr (1885–1962) and Werner Heisenberg (1901–1976), tends to privilege the methodological-epistemic layer and the operational form of what can be said under conditions of measurement, practising ontological prudence by refusing speculation about underlying realities. In the Many-Worlds interpretation, proposed by Hugh Everett III (1930–1982), probability is pressed into functioning as a measure over branches, shifting the burden to the semantic-ontological layer. In hidden-variable theories, such as that of David Bohm (1917–1992), there is an impulse to reabsorb randomness as epistemic ignorance, shifting the problem to the delimitation of what counts as a complete state. In radical Bayesian programmes that re-anchor probability as the agent's credence, the traditional direction of justification is reversed, making internal coherence the dominant axis. In objective-collapse theories, additional physical mechanisms are introduced that increase the weight of experimental testability at the price of explicit ontological commitment.

The central slippage punished by this domain is the clandestine mixing of layers. In the vocabulary proposed in this book, we shall call this error the Collapsed-Layer Fallacy. It is a category error: a confusion among types of claims and a movement from one layer to another without a declared bridge. The diagnostic question should always be asked in the same form. Does this sentence purport to describe a process in the world, prescribe an inferential rule under finite evidence, or certify a criterion of intersubjective control of the experiment? If the answer changes midway through the argument without declaration, the collapse has already occurred.

9.4 Statistical mechanics

Here the criterion of radical transparency from Chapter 8 is placed under particular pressure because delimitation and reference measure decide what counts as neutrality and what counts as 'uniform'. If in quantum physics the central epistemological challenge is to separate layers in a field with precise formalism but disputed interpretation, in statistical mechanics the axis of the problem shifts. The primary question is not what probability represents, but how to legitimate emergent descriptions when multiple delimitations are possible. Here objectivity depends less on the realism of chances than on the auditability of descriptive choices.

In statistical mechanics, probability operates as a bridge between microdynamics and macroscopic regularities when detail is intractable and collective phenomena are robust. In the Gibbsian formulation, an ensemble can be presented heuristically and historically as an imaginary collection of independent copies of the same type of system, subject to the same macroscopic constraints and distributed over different microstates. Formally, this construction is represented by a probability distribution over the state space. The collection should not be reified as a physical population existing in space-time; nor is it incorrect to use it as a formal device for interpreting ensemble averages. The semantic-ontological layer appears when we ask what this distribution says about the actual system; the methodological-epistemic layer governs the choice of ensemble and constraints; and intersubjective control requires that the delimitation and reference measure be open to criticism.

The conflict goes beyond merely expository interest. It is structurally revealing because it makes explicit that objectivity can mean, depending on the layer, impersonal rationality or publicly verifiable performance, and that the two coincide only when the regime, delimitation and reference measure are exposed and supported by results. A recurrent slippage appears when maximum entropy is presented as neutrality in itself. In the vocabulary proposed in this book, we shall call this error the Neutral-Entropy Fallacy. It occurs when what depends on the choice of description space, the reference measure that defines uniformity and the constraints treated as informative is itself treated as 'neutral'. The diagnostic question is simple and non-negotiable: which reference measure is defining uniformity, and which constraints have been treated as informative?

Another slippage appears when the ensemble construction is converted, without additional argument, into a literally existing physical population. In the vocabulary proposed in this book, we shall call this error the Reified-Ensemble Fallacy. The fault does not lie in representing the ensemble as an imaginary collection of copies—a traditional idiom since Gibbs—nor in treating it formally as a distribution over states. It arises when the ontology of a representational device is transferred to the world without declaring the bridge that warrants that passage.

9.5 Population genetics

Here the criterion of epistemic regime from Chapter 8 is placed under particular pressure because inference is almost always indirect, under rival models, and counterfactuals enter as a constitutive component of the number. In population genetics, probability circulates across levels of description that require categorical distinction. There is process randomness, parametric uncertainty and model uncertainty. The methodological-epistemic layer is dominant because almost everything is inference under indirect evidence and rival models. The semantic-ontological layer appears as a constant temptation when genes and alleles are discussed as though numbers were bare properties. The intersubjective-control layer demands radical transparency about choices, data, parametrisations and counterfactuals. Confusing these levels is the most direct route to making a number appear to be an intrinsic attribute of what is, in fact, a relational assignment under counterfactuals and stability assumptions.

Read through the analytical scheme proposed in this book, the historical contrasts among Ronald Aylmer Fisher (1890–1962), Sewall Wright (1889–1988) and John Burdon Sanderson Haldane (1892–1964) help to illustrate the problem, but they should not be taken as these authors' own self-descriptions. The Fisherian tradition emphasises design, inference and public control of procedures; Wright assigns a central role to stochastic processes, such as drift, in the study of evolution; Haldane combines mathematical modelling with substantive questions of selection. The aim here is to show how different modelling choices activate different layers of interpretation, without reducing the history of population genetics to the vocabulary of pluralism developed in this book.

The characteristic slippage here is to convert model dependence and counterfactuals into a semantics of essence. In the vocabulary proposed in this book, we shall call this error the Substantialised-Fixation Fallacy. It occurs when a relational number, defined under conditions, counterfactuals and a stability regime, is treated as though it were an intrinsic attribute of the allele. The diagnostic question is unavoidable: which assumptions and which counterfactuals support the calculation, and have they been replaced by a semantics of essence?

9.6 Epidemiology

Here the criterion of purpose from Chapter 8 is placed under particular pressure because the same number circulates among public policy, clinical practice and social communication, and moral cost appears when scales are confused. If population genetics puts pressure on the bridge between model and world through counterfactuals and modelling choices, epidemiology shifts the tension towards purpose, because the same number can operate as a tool of public policy, a clinical criterion and a social message. The passage from one use to another is precisely where semantic slippages acquire moral cost.

In epidemiology, the core methodological issue is purpose and its moral cost. The methodological-epistemic layer governs the construction of risks under data and models. The intersubjective-control layer sustains public trust through design, bias control, calibration and reproducibility. The semantic-ontological layer becomes dangerous when a number is imported as an internal chance of the person, as though the world had inscribed in the individual what was constructed through population comparison.

Frequentism and classical inference sustain public trust through design, bias control, quantification of uncertainty and institutional auditability. Bayes enters as a technology for co-ordinating heterogeneous evidence, but objectivity is preserved only when the inferential path is exposed through initial probabilities, model, calibration, sensitivity analysis and predictive assessment. Clinical traditions tend to privilege inference and decision and therefore encounter the criterion of purpose at its most sensitive point, because singular decisions require integrating evidence without promising the individual what only the population guarantees. Social and critical traditions insist that categorical delimitation is not neutral, shifting the focus towards transparency and public justification of the semantic basis that precedes calculation.

A particularly serious slippage is to convert a population measure directly into an intrinsic causal property of the person. In the vocabulary of this book, we shall call this error the Individualised-Propensity Fallacy. It occurs when a frequency, population risk or predictive probability is presented, without additional modelling and justification, as though it were an ontological chance inscribed in the patient. This does not mean that individual probabilities are illegitimate: models can produce individualised predictions, and causal theories can support claims about responses under intervention. The diagnostic question is different. Is the number cited a population rate, an individual prediction conditional on covariates and a model, or a claim about the individual's causal propensity? The transition among these readings must be made explicit.

9.7 Economics and finance

Here the criterion of regime from Chapter 8 is placed under particular pressure because reflexivity can erode the generating process itself and dissolve the basis for long-run validation. If epidemiology exposes the ethical cost of semantic transposition, economics and finance expose its structural epistemic cost. Here the problem is not merely misinterpreting a number, but presupposing that a sufficiently stable regime exists for the number to play the role assigned to it.

In economics and finance, the criterion of regime is the arena of epistemic conflict. The intersubjective-control layer may be technically rigorous and yet insufficient, because the greater difficulty is not merely measurement or estimation error but instability of the generating process under reflexivity. The methodological-epistemic layer may be sophisticated and yet fail when applied outside its conditions of validity. The semantic-ontological layer is tempting when contingent regularity is converted into law, as though the system were insensitive to the circulation of models and expectations.

This domain reveals the structural limitations of naive probabilism for a constitutive reason. Systems change with rules, incentives, institutions and expectations, and they often also change because the model itself becomes a common belief and begins to guide actions. Reflexive non-stationarity is an instance of maximal tension. It is not merely parameter change under policies; it is the mechanism through which a model, by becoming a common belief and guiding decisions, can alter the reality it describes, becoming a self-fulfilling or self-defeating prophecy and dissolving the basis of its own predictive validation. Without a minimally sustainable regime of phenomenal stability, long-run guarantees lose their ground and refinements may merely sophisticate the illusion.

The characteristic slippage is to turn historical contingency into natural law. In the vocabulary proposed in this book, we shall call this error the Structural-Parameter Fallacy. It occurs when modality is projected improperly, treating parameters estimated under a specific historical regime as invariant properties of the system while ignoring that their stability depends on institutions, expectations and policies that may change reflexively. The diagnostic question is demanding: which stability assumption supports the generalisation, over what horizon, under which mechanisms of rupture, and through what routes might the model itself become a causal agent in the system it purports to describe?

9.8 Social sciences

Here the criterion of radical transparency from Chapter 8 is placed under particular pressure because categorical delimitation founds the space of possibilities and decides what will count as evidence before calculation. If economics and finance show that regimes can be eroded by reflexivity, the social sciences show that the delimitation itself—the choice of categories and units—can carry commitments that precede calculation and condition what will be counted as evidence. Radical transparency therefore moves to the initial stage of the inferential path.

In the social sciences, radical transparency shifts to the foundational stage of the methodological process. The intersubjective-control layer is indispensable when design, sampling and measures are well defined, but this is not enough when categorical delimitation decides what counts as a variable, unit and comparison. The semantic-ontological layer is especially vulnerable because categories may be naturalised and treated as though they were givens of the world. The methodological-epistemic layer can uncritically inherit delimitations that already carry normative and historical commitments.

Frequentism is indispensable as a discipline of error and methodological publicity when design and sampling are well defined. Bayesianism is useful for decision under uncertainty and heterogeneity, provided that the delimitation is made explicit. Positivist traditions tend to privilege audit and repeatability, but they may conceal the categorical delimitation that decides what counts as a variable and as a unit. Interpretive traditions associated with Max Weber (1864–1920) insist that categories carry meaning and that the semantic layer is not a technical detail. Critical traditions require delimitation to be treated as a constitutive component of the object of study rather than a mere convention, because that is where naturalisation takes hold.

The characteristic slippage is to naturalise categories and, consequently, to call the numerical result objective as though objectivity were a property of the number rather than of the path. In the vocabulary proposed in this book, we shall call this error the Naturalised-Category Fallacy. It occurs when the category that defines the space of possibilities is treated as though it were a natural and neutral delimitation and, from there, the numerical result is reified as a neutral photograph of reality. The diagnostic question must be asked before any estimation: which categories define the space of possibilities, what do they exclude, and what public justification supports their adoption in this context?

9.9 Limiting case

If the social sciences require criticism of the delimitation to precede calculation, the limiting case shows where even that precaution is insufficient because the kind of public reference that would permit probabilities to be interpreted as chances in the world or as long-run guarantees is absent. Probability must be explicitly downgraded to a heuristic, lest it become rhetorical authority.

The limiting case protects pluralism from an imperial temptation. There are domains in which probability should not be sold as chance in the world or as a long-run guarantee, because what exists is comparison among scenarios under disputable premises and fragile counterfactuals. Assigning probabilities to the fall of the Roman Empire or to the emergence of life on Earth depends on post hoc delimitations, debatable counterfactuals and highly underdetermined fundamental models. Whereas quantum physics has robust experimental audit and epidemiology has public reference in trials and cohorts, here both are lacking. The number, if used, functions as a heuristic tool for organising plausibilities under declared premises. Responsible use requires explicit declaration of what the number presupposes and what it does not purport to assert, so that auditability of the path is not replaced by rhetorical authority. The limiting case, therefore, is not an exception but the stress test of pluralism. It shows where the architecture of criteria remains standing precisely by refusing the pretence of providing ontological answers where only useful heuristics are available. A second example helps to fix the point. In counterfactual history or in the palaeontology of rare events, such as probabilistic reconstruction of causes of extinctions, the number organises plausibilities under premises and underdetermined models, but it does not, by itself, purchase the status of chance in the world or long-run guarantee.

9.10 Anticipating objections: limits and responses of pluralism

A practical objection would say that the protocol is too complex for everyday application. The answer is that it is not intended to replace working heuristics, but to provide a standard of public justification when results are disputed, when inferences cross domains and, above all, when consequences are serious. The protocol is closer to a discipline of conceptual audit than to a manual of calculation.

A second objection would charge relativism. If there is no ultimate foundation, everything would be convention. The answer is that the pluralism defended here is anti-foundationalist but not relativist. It requires conventions to be declared and auditable, and it transforms metaphysical disputes into debates about methodological adequacy, transparency and conditions of application. Instead of 'anything goes', the requirement is 'nothing goes without a publicly justifiable path'.

A third, imperial objection would ask why everything should not be reduced to credences and conditionalisation, taking Bayesianism as a universal metalanguage. The answer emerges clearly from the analysis of the domains. The intersubjective-control layer requires a publicity of procedure involving design, calibration, reproducibility and peer criticism that the internal coherence of a system of credences does not, by itself, provide. Programmes that radicalise probability as credence may function as philosophical experiments, but clinical epidemiology and population genetics cannot renounce the public control that grounds their social credibility. The protocol is not anti-Bayesian. It is post-imperial, refusing to dissolve the functional distinctions that make the most reliable epistemic practices intelligible.

A final objection would charge conservatism. By requiring transparency and declaration of assumptions, the protocol would freeze innovation. The answer is the reverse. By making delimitations, layers and regimes explicit, the protocol shows precisely where innovation is breaking previous assumptions, thereby making the point of rupture clearer and the act of replacing one regime with another more responsible. Scientific revolutions are not impeded by explicitness; they are strengthened by it, because they cease to depend on rhetoric and come to depend on justifications that can be criticised and improved.

9.11 Synthesis

The synthesis consolidates the conceptual advance of the chapter in two complementary operations. The first is to make explicit the logical cascade of epistemic dependence. Delimitation defines the space of possibilities; the layer defines the type of claim; the regime determines whether there is a basis for generalisation; reflexive consistency prevents semantic smuggling between justification and use; radical transparency makes the entire path reconstructible by competent peers. Each domain studied places pressure on one particular link in this cascade and exemplifies the primacy of one criterion. Quantum physics makes the criterion of layer separation unavoidable. Statistical mechanics demonstrates the cost of violating the criterion that delimitation and reference measure be made explicit. Genetics reveals the danger of neglecting the criterion distinguishing model from world and of converting counterfactual relations into essential attributes. Epidemiology starkly exposes the criterion of purpose and the ethical cost of violating it. Economics and finance place at the centre the most challenging criterion: stability of the regime under reflexivity. The social sciences require the criterion of radical transparency to extend to the foundational categorical vocabulary.

The second operation is to make explicit the logical architecture of the fallacies as a taxonomy of categorically improper projections and to record their kinship with familiar distinctions in philosophy of language. The Collapsed-Layer Fallacy is the general form of category error, through confusion among logical types of claims. The Reified-Ensemble Fallacy projects ontology where there is a formal construct. The Individualised-Propensity Fallacy projects an individual scale where there is only a population measure. The Structural-Parameter Fallacy projects necessity where there is historical contingency dependent on regime. The Naturalised-Category Fallacy projects a natural given where there is historical-social construction. The Neutral-Entropy Fallacy projects neutrality where there is a convention of measure and description. The Substantialised-Fixation Fallacy projects inherence where there is a counterfactual relation under a model.

Philosophically, this taxonomy echoes classic distinctions about conditions of conceptual application. The Collapsed-Layer Fallacy corresponds to a category mistake, in a sense close to Gilbert Ryle's (1900–1976) diagnosis, because it confuses logical types of claims. Projection fallacies, such as Individualised Propensity, Structural Parameter and Naturalised Category, violate what Ludwig Wittgenstein (1889–1951) would call the grammar of probabilistic concepts, applying them outside their conditions of use. Reification fallacies, such as the Reified Statistical Ensemble and Substantialised Fixation, approach Wilfrid Sellars's (1912–1989) criticism of the myth of the given by turning epistemic decisions and methodological constructs into supposed ontological facts. The protocol that follows functions as a methodological antidote to these conceptual pathologies, not by eliminating substantive disputes but by preventing them from being installed clandestinely.

The protocol therefore operates as a cascade of checks that must be read as a sequence of cumulative requirements. Stage 1: configuration of the inferential space. The domain of application must be delimited, with variables, categories and exclusions declared, and the assertoric status of the probabilistic claim must be defined, clarifying whether it is intended to operate in the semantic-ontological layer, the methodological-epistemic layer or the intersubjective-control layer. Stage 2: verification of conditions of validity. The analysis bifurcates because regime has two faces that must not be allowed to oscillate. The regime of phenomenal stability must be declared, with an assumption about the generating process, temporal horizon of validity and conditions of rupture; and the regime of methodological validity must be declared, justifying why the inferential rule adopted is appropriate to this delimitation and purpose and under which conditions of application it is legitimate. Stage 3: final controls. Reflexive consistency is required so that the interpretation of the result remains governed by the justification of the calculation and so that clandestine transpositions among semantics are blocked; and radical transparency is required, keeping initial probabilities, reference measures, parametrisations, model choices, validation criteria and sensitivity analyses publicly available.

To show the protocol in action, it is useful to apply it to a simple statement in quantitative linguistics. Consider the claim 'the probability that a word in Portuguese is a preposition is 12%'. Under delimitation, the initial question decides almost everything. Which corpus defines 'Portuguese'? Does it include speech and writing? From which period and which genres? By which tokenisation criteria is it decided what counts as a word, especially in contractions and hyphenations? Under the layer criterion, it must be stated whether this is a methodological-epistemic claim—that is, an observed frequency under a delimitation—or whether it purports to be a semantic-ontological assertion about a propensity of the language, as though the language possessed its own rate independently of delimitations and contexts. Under the regime of phenomenal stability, one asks whether diachronic stability or stability across genres is being assumed and how the proportion of prepositions behaves under historical and sociolinguistic variation. Under the regime of methodological validity, one asks why a frequentist count is appropriate, whether the size and composition of the corpus support convergence within the delimitation, and exactly what is intended to be inferred from the figure. Under reflexive consistency, the rule is imposed not to interpret as a structural law of the language what has been justified only as a frequency under a delimitation, unless new justification is provided. And under radical transparency, the corpus, tokenisation criteria, exclusions and sensitivity to alternative choices must be publicly available so that the reader can reconstruct the path. Only after this process does the statement acquire inferential legitimacy, and even then only for the declared delimitation.

The final effect is the intended one. Probability ceases to be a password and becomes grammar; grammar ceases to be abstract theory and becomes an intersubjectively auditable practice of epistemic responsibility for the inferential path.

[1] In the literature, this family of positions is known as QBism (Quantum Bayesianism).

[2] In the literature, the most frequently cited objective-collapse theory is GRW (Ghirardi-Rimini-Weber).

10 Conclusions and Horizons in the Twenty-First Century

(Probability as public infrastructure)

This book has traversed a territory in which 'probability' often functions as a high-powered label, an intellectual currency that purchases credibility. The same word opens different doors, legitimates different inferences and sustains decisions with asymmetric costs. The journey through the preceding chapters revealed a pattern that is not accidental. Each interpretation is a particular infrastructure that enables certain paths and blocks others. When an interpretation promises objectivity, it installs it in a specific place and, in doing so, accepts a price elsewhere. The gain is real. So is the cost.

The conclusion, therefore, is not a verdict naming an overall winner. It is a theory of forced choices and an ethic of use suited to a century in which probabilistic numbers circulate as public infrastructure. What is required at the end is not the search for a single semantics, but the discipline of declaring what one is doing when the same algebra is used for different functions. In the twenty-first century, semantic failure is institutional failure.

10.1 Map and costs of the interpretations

The classical interpretation aspired to objectivity in the pure form of rational symmetry. Its brilliance appears when the space of possibilities is stabilised by robust practices, instruments and conventions. Its cost emerges when that stabilisation is absent. The problem is not the algebra. It is the choice of space and measure. Joseph Bertrand's (1822–1900) paradox made the blind spot explicit. When the classical interpretation speaks of equipossibility, it depends on prior structures of individuation and measure. The promise of neutrality does not arise from nothing. It arises from a set-up.

Frequentism aspired to objectivity in public regularity. Its advantage was to turn probability into an auditable guarantee of procedure rather than a quality of the case. The technology of public control consolidated in the twentieth century by Ronald Fisher (1890–1962), Jerzy Neyman (1894–1981) and Egon Pearson (1895–1980) offered an operational ethic. This guarantee is defined in terms of the behaviour of the procedure under repeated sampling, actual or idealised, with goals such as error control and coverage. The cost of this advantage was to make the singular case legible only through the mediation of the population. Frequentism is impeccable in repetition and struggles when the world demands direct judgement about a unique event, because its objectivity is, by definition, a property of series.

Propensity theory sought to reinstall objectivity in the world. Karl Popper (1902–1994) sought to anchor chances in the experimental set-up and thereby save the singular case without appealing to the epistemic agent. The gain is attractive in domains where the language of dispositions seems to capture a real physical feature. Its price proved to be a constitutive tension. Causal ontology is asymmetric. The algebra of inference is symmetric. The two are not co-ordinated without a principle that prevents causal direction from being confused with inferential direction. Testing a propensity requires regimes of repetition or, at least, generating conditions sufficiently stabilised for intersubjective comparison and co-ordination with evidence. Paul W. Humphreys's (1950–2022) criticism makes visible the friction between 'chance in the world' and 'inverse probability' as a tool of inference. An ontology of chances does not, by itself, become a complete semantics for every use of the calculus.

Best-system interpretations occupy another position on this map. They too preserve objective chances, but seek to reduce them to the probabilistic structure of the system of laws that best balances simplicity, strength and fit with the history of the world. In this way, they avoid identifying chance with a local power of the set-up, but they make objectivity dependent on the global assessment of a systematisation. The gain is a more parsimonious ontology; the cost is justifying the criteria that make one system better than its competitors (Lewis, 1994; Hoefer, 2019).

Logicism aspired to objectivity in impersonal validity. John Maynard Keynes (1883–1946) sought intersubjective confirmation as a rational relation between propositions. Rudolf Carnap (1891–1970) attempted to provide a mechanism through formal languages and confirmation functions. The gain was a conceptual discipline that compels assumptions of language and evidence to be made explicit. The cost was dependence on conventions and intractability. Willard Van Orman Quine's (1908–2000) criticism displaced the foundational ambition by showing its dependence on a web of belief. Logicism did not disappear. Its status changed. It ceased to be an ultimate foundation and remained as a regulative ideal and intellectual hygiene.

Bayesian subjectivism aspired to objectivity through discipline of the subject. Frank P. Ramsey (1903–1930) showed that degree of belief and preference are constitutively bound together, on pain of incoherence. Bruno de Finetti (1906–1985) radicalised the thesis and, through exchangeability, explained how practical objectivity can emerge 'as if' there were a stable parameter. The gain is direct applicability to the singular case and a rational grammar of updating under partial evidence. The cost is responsibility for the starting point and the model delimitation. There is no zero point. There is procedural integrity, transparency and openness to learning.

These attempts are not random. They expose conceptual boundaries that define the terrain on which probability can be objective. First, there is a limit of space. There is no probability without a space of possibilities, and that space always comes with language, reference class, parametrisation, base measure and model. Second, there is a limit of the singular case. Frequentism mediates it through classes. Propensity theory and subjectivism confront it directly and pay, respectively, with ontological commitment and epistemic responsibility. Third, there is the limit of stability. Every form of probabilistic objectivity presupposes some stability—physical, institutional, linguistic, inferential or model-based. When that stability is contested, the probabilistic number must carry the stability assumption as part of the argument rather than as background. Finally, there is the limit of a single semantics. The same algebra serves different layers, and abuse arises when the authority of one layer is lent, without warning, to another.

10.2 Pluralism and responsibility

Faced with this map, the easy way out is weak ecumenism: everyone chooses their interpretation and moves on. That stance dissolves conceptual costs and turns pluralism into licence. The map of gains and costs, however, is not an invitation to scepticism. It is an invitation to managerial clarity. If each interpretation installs objectivity in a specific place and pays a price elsewhere, then the choice cannot be treated as blind doctrinal adherence. It must be a justified methodological decision in which the assumed cost is compensated by adequacy to the problem, domain and purpose. This is where weak ecumenism proves insufficient and dangerous.

Criterion-guided pluralism is the opposite of this weak ecumenism. It functions as a theory of methodological decision and maintains that, given the tensions identified, managing trade-offs with public responsibility is an especially robust strategy. It does not claim that every monism is incoherent by definition; it claims that any aspiration to uniqueness must demonstrate how it preserves competing virtues without concealing the costs that the map has made visible. This requires a citable and verifiable procedure that makes methodological responsibility operational.

Let us call this procedure the Inferential Declaration Protocol. It does not arise by decree. It crystallises, into minimum duties, the conceptual architecture built throughout the book. It is the institutionalisation of the requirement of semantic consistency. The justification must govern the type of conclusion. It requires the active layer—semantic, inferential, audit or decision-making—to be declared. By the decision-making layer we mean the use of the number as an input to action under criteria of loss, value and responsibility, where objectivity resides in the rationality of the decision given a structure of preferences rather than primarily in description of the world or methodological guarantee. It requires the delimitation to be made explicit: reference class, language, model, parametrisation, base measure, constraints and relevant exclusions. It requires transpositions to be justified when a procedural guarantee is used to support a singular claim, when a predictive probability is treated as explanation without a declared mechanism, or when an epistemic probability is presented as though it were objective chance. It further requires limits of inferential validity to be declared, specifying under which conditions of stability, context and intervention the results may be generalised and which regime ruptures would tend to invalidate them. This turns the stability assumption from background into a declared frontier of the knowledge produced. Finally, it requires the path to be submitted to public criticism, documented so that it can be replicated, contested and tested.

In a more stable form, the Protocol is organised into four layers of mandatory declaration: (1) Semantics: what the number purports to say about the world; (2) Inference: how it was obtained under evidence and model; (3) Audit: how its performance is publicly controlled; (4) Decision: how it guides action under values, losses and responsibility. The golden rule remains the same: justification must govern the conclusion.

Criterion-guided pluralism, therefore, does not describe preferences. It defines responsibilities. Ignoring this protocol is not choosing a different school. It is renouncing the epistemic accountability that distinguishes science from persuasion by numbers. For example, when a calibrated score is presented as an individual 'causal probability', the authority of performance is harvested to sustain a semantics that was never declared. In the twenty-first century, this renunciation produces institutional failures and material harms because probabilistic numbers guide medicine, public policy and automated systems.

10.3 Association, mechanism, complexity

Probability is not causality. It may accompany causality, help infer it and summarise regularities produced by mechanisms. It does not replace explanation. Complex systems make this distinction even more vital because they multiply dependencies, heterogeneity and regime change.

Here the Inferential Declaration Protocol becomes more demanding. In causal research, making the delimitation explicit means more than defining a population. It means specifying the counterfactual space, which interventions are being considered, which variables are assumed fixed and which architecture of potential responses is in play. A regression coefficient is not a causal effect without design and declared counterfactual assumptions.

Likewise, justifying stability ceases to be background and becomes a defence of invariances under interventions and changes of context. In complex systems, a central part of scientific work is to investigate which relations survive perturbations and which collapse, and to report this as part of the result—that is, as part of the epistemic content of the inference.

Here the criterion of reflexive consistency operates at full force, functioning as a veto on semantic leaps. A probabilistic association, however strong and apparently stable, does not by itself warrant a causal arrow. To make that leap, one must declare assumptions, mechanisms and tests that render it plausible and controllable under intersubjective audit.

In complex systems, the mature question is rarely 'what is the probability?'. It tends to be another: under which architecture of mechanisms and invariances does this probability warrant which inference, and with what scope of validity?

10.4 Data, performance, meaning

The twenty-first century has intensified the dilemma by automating inference and multiplying data. This reconfigures the relation between performance and meaning. A model can perform brilliantly at the layer of predictive performance, including under empirical audit, and yet collapse—or become empty—at the semantic and explanatory layers. Calibration measures numerical agreement under a declared delimitation and under conditions similar to those of validation. It does not measure causal intelligibility or the robustness of the categorical vocabulary that sustains the score itself.

The challenges introduced by automation and data volume are, on a massive scale, an extension of the limiting case analysed earlier. When models operate in high dimensions with optimised predictive performance, an illusion is created that the limits of inference have been overcome, as though calibration could replace understanding. In fact, pressure on delimitation, stability and transparency becomes even greater because technical complexity can conceal semantic fragility.

In many contemporary uses, 'probability' is a score. The basic test is calibration. If a system produces 0.8, one expects events to occur at approximately that rate, on average, within defined groups and under a declared delimitation, under conditions similar to validation. This operational objectivity is valuable. It is not explanation.

A system can be calibrated and yet be causally wrong, biased in subgroups or fragile under regime change. The first pressure is the breakdown of stability. A model calibrated in one period can become unfair and inaccurate during an economic, health or institutional rupture, not necessarily because of a hidden defect in the algorithm, but because of an undeclared violation of the regime assumption. Added to this is an illusion of stability produced by volume. A model trained on a vast quantity of pre-pandemic traffic data may appear impeccable and yet become blind to the complete disruption of patterns caused by a public-health lockdown. The past ceases to guide because the world changes, and the number alone does not contain the clause of validity that made it legitimate.

The second pressure is model uncertainty. Estimating parameters is not enough. We must recognise that the model may be wrong, that omitted mechanisms matter and that proxy variables carry history, power and dependence on measurement. When this structural uncertainty is silenced, probability becomes high-dimensional numerology, with the appearance of rigour and little epistemic security.

A third pressure, aggravated by Big Data, is the illusion of completeness. Masses of data can create the appearance of a stable regime where there is only the inertia of past processes, masking break points. The declaration protocol compels the question that cannot be avoided: is this regularity a feature of the phenomenon or an artefact of the archive?

The decisive point here is social. Automated models operate on categories. Categories define spaces. Delimitation and stability cease to be philosophical jargon and come to encode robustness, fairness and legitimacy. Auditing them is not an ornament. It is a condition of public validity. This is where the fallacies of Chapter 9 inevitably return in a contemporary key. When a calibrated score is converted into personal destiny, the Individualised-Propensity Fallacy analysed in Chapter 9 is materialised at algorithmic scale, now amplified by automated systems.

10.5 Public agenda

The legacy of this journey is not a single answer. It is a tool of scrutiny and a programme of reconstruction. If probability is to remain a legitimate language of science under uncertainty, it needs bridges.

The common project of these bridges is singular: to make the passage from calculation to the world auditable.

A first bridge is semantic. It connects philosophers, statisticians and data scientists in order to consolidate a cartography of layers and uses, so that 'probability' ceases to be an ambiguous label and becomes an explicit declaration of function. The practical question is unavoidable. How should journals, textbooks and courses be redesigned so that declaration of the layer—for example, 'this is a long-run probability', 'this is a calibrated credence', 'this is an operational guarantee under audit'—becomes as standard as reporting a p-value or a confidence interval?

A second bridge is methodological. It turns delimitations into public infrastructure. Reference class, parametrisation, base measure, modelling choices and exclusion criteria must be treated as constitutive parts of the argument rather than footnotes. The concrete action is to establish reporting formats that require these choices to be visible, not as bureaucratic formalism but as a condition of critical replication.

A third bridge is scientific. It establishes habits and methods for testing, debating and reporting hypotheses of regime change, non-stationarity and inferential fragility, treating stability as a hypothesis rather than an invisible assumption. Publication standards should normalise declaration of the horizon of validity and expected mechanisms of rupture, so that 'it works under these conditions' has the same status as 'we obtained this number'.

A fourth bridge is theoretical. It confronts the problem of co-ordinating objective chance and rational degree of belief in a world of approximate models and finite evidence, without confusing causal direction with inferential direction. This requires, on the one hand, theories that explicitly represent dependence on delimitations and reference measures and, on the other, practices that make initial probabilities and updating conditions visible, so that disagreements are not masked by superficial numerical consensus. For example, hierarchical models that make dependencies among levels explicit and causal-inference frameworks that graphically represent assumptions of delimitation and conditional independences can operate as formal instruments of this transparency.

A fifth bridge is ethical and institutional. It establishes a governance of trust in which declaration of assumptions, sensitivity analysis, limits of validity and compatibility between justification and use become ordinary language of practice. One concrete action is the adoption of inferential-declaration appendices in scientific publications and technical reports, in which delimitations, stability assumptions, sensitivity analyses, validity limits and audit criteria are documented as meticulously as code and data.

Probability has never been the universal key it is sometimes promised to be. Its different concepts can describe chances, regularities, support relations or degrees of belief, depending on the regime adopted. What should be common to these uses in science is a different requirement: a public protocol for communicating and justifying uncertainty honestly. The challenge of the twenty-first century is not to abolish semantic differences, but to declare delimitations, conditions of stability, updating criteria and limits of validity, preventing the authority of one use from being clandestinely transferred to another.

The work of this book, therefore, has not been to identify one true interpretation, but to contribute to the foundations of a publicly trustworthy inferential practice. In this practice, probabilistic numbers do not circulate as currency of autonomous authority, but as auditable components of collective reasoning. Under this regime, the crucial question will cease to be 'which interpretation do you subscribe to?' and will become another: are your criteria declared? Is your protocol public? Is your bridge between justification and consequence legitimate? This is the standard that can transform probability from a label of authority into an infrastructure of trust, a language worthy of a century decided by data but defined by the clarity of those who construct, audit and use them. This is the possible legacy of the journey proposed by this book: not a final answer, but an invitation to public responsibility for the numbers with which we decide the future. Probability is public only when the path is public—and it is worthy of trust only when the path is declared.

11 Glossary

Almost-sure convergence: Convergence of random variables that occurs on a set of probability 1. It is equivalent to $X_n(\omega) \to X(\omega)$ for almost every ω, in the sense of P.

Association: Statistical relation between observed variables, expressed through dependencies, correlations or effects estimated in models. It does not imply causality without a bridge of intervention and causal delimitation.

Auditability: Capacity of a probabilistic result to be verified by third parties through an explicit protocol, data, design and criteria. It is a methodological, not metaphysical, virtue.

Bayes (Bayes' rule): Updating rule that transforms initial probabilities into posterior probabilities by combining model and data. It requires explicit declaration of what counts as hypothesis, datum and delimitation.

Bertrand paradox: Example showing that 'at random' in continuous settings depends on the procedure and measure. It serves as a conceptual microscope for delimitation.

Best-system account: Analysis of objective chance in which chances are the probabilities determined by the probabilistic laws of the system that best balances simplicity, strength and fit to the history of the world. It differs from frequency and propensity accounts.

Calibration (predictive): Property by which predictive probabilities coincide with observed frequencies in comparable classes. A public criterion of predictive quality.

Causal inference: Inference about effects of interventions, based on counterfactuals and causal assumptions. It is not reducible to observational conditionals.

Causality: Relation of production or intervention linking actions to effects and permitting counterfactual inference. It is not reducible to observational association.

Central limit theorem: Family of results in which sums or means of random variables, under stated conditions, approach a Normal distribution after standardisation. It justifies the ubiquity of Gaussian approximations and should be applied with explicit assumptions.

Classical probability: Probability as the ratio of favourable to possible cases under equipossibility. It requires stabilised symmetry and delimitation.

Coherence: Normative constraint on degree of belief that avoids practical contradictions, such as guaranteed losses in betting. It is not synonymous with 'statistical consistency'.

Collective phenomena: Domain in which regularities emerge in populations, series and repetitions under stable conditions, making long-run guarantees viable. In this book, this is the canonical term used to avoid confusion with von Mises's 'collective'.

Collective (von Mises): Idealised sequence in which frequencies converge and remain invariant under admissible selection rules. It is intended to ground the frequentist idea of probability as a limit.

Conditional probability: Probability of one event given another. It should not be confused with interventional probability in causal analysis.

Conditioning: Operation of updating probabilities by information, that is, through conditional probability. It should not be confused with causal intervention.

Confidence interval: Procedure that, under repetitions, covers the parameter at the nominal frequency under the model. It does not express degree of belief without an explicit transposition.

Confirmation: Degree of evidential support that evidence confers on a hypothesis within an inferential framework. It is not 'chance in the world' and is not, by itself, frequency.

Consistency (formal): Property of a system whereby it avoids internal contradictions. In this book, it refers to the formalism and rules of calculation, not to beliefs.

Consistency (statistical): Property of an estimator whereby it converges to the target as sample size increases under stated conditions. It should not be confused with coherence of degree of belief.

Convergence in probability: Mode of convergence of random variables in which, for every $\varepsilon>0$, $P(|X_n-X|>\varepsilon)\to 0$ as $n\to\infty$. It captures probabilistic approximation without requiring pointwise convergence.

Counterfactual: Statement about what would have occurred under a different intervention. A conceptual basis of causal inference.

Counting measure: Natural reference measure on discrete spaces, assigning to a set the number of its elements (and ∞ when infinite). It serves as a basis for equipossibility by counting.

Credence: Quantitatively represented degree of belief. In this book, the term is used when the technical and normative sense is to be emphasised, distinguishing it from merely descriptive psychological belief.

Decision: Choice of action under uncertainty, often modelled through expected utility. It requires values, costs and the decision-making layer to be declared.

Degree of belief: Graded inferential commitment of the agent, regulated by norms of coherence and updating. It differs from emotion and from objective chance.

Delimitation: Specification of the domain, hypotheses, language, parametrisation, measure and objectives. A condition of legitimacy for assigning probability.

Dependence: Statistical relation in which events or variables are not independent under the model. It does not imply causality.

Design (experimental): Set of rules for data collection and allocation that grounds inference. An essential part of delimitation.

Distribution (law): Probabilistic description of the outcomes of a random variable in the model. It can be used in different layers, but its delimitation should be declared.

Ensemble (statistical mechanics): Formal construction that can be presented as an imaginary collection of independent copies of a system under the same macroscopic constraints and is represented by a probability distribution over possible states. By itself, it is not a physical population of actually existing copies.

Entropy: Measure of the dispersion or information of a distribution, used in criteria such as maximum entropy. It depends on the constraints chosen.

Epistemic accuracy: Criterion that evaluates credences by their closeness to truth, usually through scoring rules. In accuracy-first epistemology, it can ground probabilistic norms without reducing their justification to betting or practical utility.

Epistemic probability: Probability as rational degree of belief, confirmation or evidential support. It depends on the informational state and norms of coherence.

Equipossibility: Assumption of equal probabilities for elementary cases under a delimitation and reference measure. In continuous settings, it requires caution and specification.

Error control: Set of long-run guarantees of inferential procedures, such as false-positive rates. It does not express degree of belief in the singular case.

Error policy: Set of conventions and long-run guarantees for evaluating procedures, for example false-positive rates and coverage. It is distinct from a measure of evidence in the singular case and from degree of belief.

Event: Measurable subset of Ω belonging to $\mathcal{F}$, to which P assigns probability. It is the formal object on which the theory operates, not a 'fact' in itself without a delimitation.

Evidence: Empirical information used to support inference. It should be linked to an updating mechanism and a delimitation.

Evidential support: Way of describing confirmation without confusing it with chance in the world. It should always be linked to an inferential framework.

Exchangeability: Assumption that the order of observations does not alter their joint distribution in the model. It is not synonymous with independence.

Expectation (expected value): Probabilistic mean of a random variable under the model measure P. Formally, when integrable, $E[X]=\int X\,dP$ (or a sum in the discrete case). It can serve decision-making when coupled to a utility function.

Framework: Set of rules, criteria and objectives determining how probabilities will be assigned, updated and used. It includes choices of delimitation, a notion of evidence, error policy and the bridge between formal number and interpretation.

Frequency limit: Value towards which relative frequency would tend as the number of repetitions grows indefinitely. A central idealisation of frequentism.

Frequentism: Interpretation linking probability to the behaviour of frequencies in series under fixed conditions. Strong in methodological publicity, fragile in the singular case without a reference class.

Frequentist probability: Probability as a frequency limit under repetitions in identical conditions. It requires a reference class and stability.

Generating mechanism: Set of conditions and procedures that produces data. A basis for discussing stability, objective chance and methodological publicity.

Guarantee (long-run guarantee): Property of a procedure under repetition, such as nominal coverage or false-positive control. It is not a promise of correctness in a unique case.

Imprecise probability: Representation of uncertainty that permits sets of distributions, intervals or lower and upper probabilities, avoiding the imposition of a single point probability when the available information does not justify it.

Independence: Condition modelled through factorisation of probabilities. In the world, it usually represents the absence of a relevant connection under the delimitation, not a metaphysical guarantee.

Indifference (principle): Rule for assigning equal probabilities when there is no declared reason to prefer one case over another. It depends on representation and reference measure.

Inference: Process of moving from data and assumptions to conclusions according to declared rules. It requires the epistemic, methodological and ontological layers to be distinguished.

Initial probabilities: Initial assignments that precede evidence and permit conditioning, inference and decision. They should be declared and justified.

Jeffreys-Lindley paradox: Phenomenon in which, for large samples, a small p-value can coexist with strong Bayesian support for the null hypothesis under certain choices of initial probabilities. It exposes a structural divergence between error policies and measures of evidence.

Kolmogorov axioms: Set of axioms characterising probability as a measure P on (Ω, $\mathcal{F}$), with non-negativity, normalisation $P(\Omega) = 1$ and σ-additivity. They provide a common syntax for the interpretations without deciding their semantics.

Law of large numbers: Result linking relative frequencies to probabilities under stated conditions. By itself, it does not solve the problem of the singular case.

Layer: Level at which probability operates in a passage, which may be formal, epistemic, ontological, methodological or decision-making. It should be declared whenever there is a risk of confusion.

Lebesgue measure: Natural reference measure on $\mathbb{R}$^n, invariant under translations, generalising length, area and volume. It is frequently used to define densities and probabilities in continuous models.

Likelihood: Function measuring the compatibility of the data with parameter values in the model. It is not the 'probability of the parameter' on a frequentist reading.

Logical probability: Probability as a support relation between propositions, dependent on language and rules. It must be distinguished from degree of belief and objective chance.

Mathematical expectation: Synonym for expected value or expectation. When integrable, E[X]=$\int$ X dP (or a sum in the discrete case), and it expresses the probabilistic mean in the model.

Maximum entropy: Principle of choosing the distribution that maximises entropy under known constraints. It represents disciplined ignorance, not absolute neutrality.

Model: Construction linking formalism to observations and interpretations. It requires specification of delimitation, parametrisation and reference measure.

Objective chance: Ontological reading in which probabilities are features of the world's set-up, for example dispositions of the generating mechanism. By itself, it does not determine degree of belief without a normative bridge.

Operational criterion: Rule that makes it possible to decide, measure or test a term within the delimitation. Example: how 'at random' is implemented, or how stability is verified.

p-value: Probability, under the reference hypothesis of the test and under repetition of the procedure, of observing a result at least as extreme as that observed according to a defined statistic. It is not the probability that the hypothesis is true and does not, by itself, measure effect size.

Parameter: Quantity indexing a family of models and organising inference. It should be interpreted cautiously because it does not always have direct ontological reality.

Partition (refinement): Division of the reference class by relevant properties in search of homogeneity. It is never guaranteed to be 'final'.

Population: Target set to which inferences are intended to generalise. It requires a clear inclusion rule and a connection to sampling.

Posterior: Abbreviated form of 'posterior distribution'. In this book, it should always be understood with the model and initial probabilities either implicit or declared.

Power: Probability, under a specified alternative and a fixed decision rule, of obtaining rejection. It depends on effect size, variance, design, sample size and significance level.

Practical limit: Notion of convergence sufficient for empirical use without literal appeal to infinity. It requires criteria of sufficiency.

Prediction: Probabilistic assignment to future data under a model. It is not a causal explanation without a mechanism.

Predictive probability: Probability of future observations given evidence and a model. It can be used as a criterion of adequacy.

Probability measure: Function P: $\mathcal{F} \rightarrow [0,1]$ defined on a σ-algebra $\mathcal{F}$ over Ω, with $P(\Omega) = 1$ and σ-additivity. It is the formal object identified with 'probability' by the axiomatic approach and requires a semantic reading for application.

Probability space: Triple $(\Omega, \mathcal{F}, P)$, where Ω is the sample space, $\mathcal{F}$ is a σ-algebra of events and P is a probability measure. It is the minimal formal structure of the axiomatic approach.

Propensity: Ontological interpretation in which probabilities are dispositions of the generating mechanism to produce outcomes. It helps to treat singular cases but requires a normative bridge.

Propensity score: Probability of receiving treatment given covariates, used for balancing in observational studies. It should not be confused with ontological 'propensity'.

Publicity: Requirement that inferences be defensible by external, auditable and comparable criteria. A characteristic virtue of frequentism and of protocols.

Random variable: Measurable function $X:\Omega \rightarrow \mathbb{R}$ (or into another space) assigning a value in the model to each outcome ω. Its distribution is the typical inferential object, not the isolated realisations.

Reference class: Set of repeatable cases within which the singular case is placed in order to receive a frequentist probability. It always requires public justification of similarity and stability.

Reference measure: Measure chosen as the basis for equipossibility, sampling or a generating mechanism. It prevents the error of treating 'at random' as an operator without a mechanism.

Relative frequency: Observed proportion of occurrences in repetitions. It is not automatically probability without an interpretation.

Sample space: Set Ω of possible outcomes of the experiment in the model. Its choice is part of the delimitation and conditions the interpretation of events and probabilities.

Semantic transposition: Passage from a formal number to a reading about the world, belief or action. In this book, it should be declared whenever there is a shift of layer.

Significance level: Parameter of error policy in frequentist tests. It is not, by itself, a measure of evidence.

Singular case: Event treated as unique, with no effective observational series available. It is the classic point of tension among interpretations.

Stability regime: Set of conditions under which the generating mechanism does not change in a relevant way. A precondition for generalisation and publicity.

Trade-off: Situation in which gaining one virtue—for example, methodological publicity—entails paying a cost—for example, difficulty with the singular case. It should be made explicit.

Weak law of large numbers: Result according to which, under standard conditions, the sample mean converges in probability to the expectation. It is a formal bridge between probability in the model and observable regularity in series.

12 References

ALDOUS, David J. Exchangeability and related topics. In: ÉCOLE D'ÉTÉ DE PROBABILITÉS DE SAINT-FLOUR XIII. Lecture Notes in Mathematics. Berlin: Springer, 1985. v. 1117, p. 1-198.

ANSCOMBE, Francis J.; AUMANN, Robert J. A definition of subjective probability. The Annals of Mathematical Statistics, v. 34, n. 1, p. 199-205, 1963.

BAROCAS, Solon; HARDT, Moritz; NARAYANAN, Arvind. Fairness and machine learning: limitations and opportunities. Cambridge, MA: The MIT Press, 2023. Available at: https://fairmlbook.org/ Accessed: 3 Feb. 2026.

BASSI, Angelo et al. Models of wave-function collapse, underlying theories, and experimental tests. Reviews of Modern Physics, College Park, v. 85, n. 2, p. 471-527, 2013. DOI: 10.1103/RevModPhys.85.471.

BERGER, James O. Statistical Decision Theory and Bayesian Analysis. 2. ed. New York: Springer, 1985.

BERNARDO, José M.; SMITH, Adrian F. M. Bayesian Theory. Chichester: Wiley, 1994.

BERNOULLI, Jacob. Ars Conjectandi. Basel: Thurneysen, 1713.

BERTRAND, Joseph. Calcul des probabilités. Paris: Gauthier-Villars, 1889.

BILLINGSLEY, Patrick. Probability and Measure. 3. ed. New York: Wiley, 1995.

BISHOP, Christopher M. Pattern Recognition and Machine Learning. New York: Springer, 2006.

BLACKWELL, David; DUBINS, Lester. Merging of opinions with increasing information. The Annals of Mathematical Statistics, v. 33, n. 3, p. 882-886, 1962.

BOHR, Niels. Atomic Theory and the Description of Nature. Cambridge: Cambridge University Press, 1934.

BOLTZMANN, Ludwig. Lectures on Gas Theory. Berkeley: University of California Press, 1964.

BOREL, Émile. Éléments de la théorie des probabilités. Paris: Hermann, 1909.

BORN, Max. Natural Philosophy of Cause and Chance. Oxford: Clarendon Press, 1949.

CARNAP, Rudolf. Inductive Logic and Rational Decisions. In: CARNAP, Rudolf; JEFFREY, Richard C. (eds.). Studies in Inductive Logic and Probability. Berkeley: University of California Press, 1971. v. 1.

CARNAP, Rudolf. Logical Foundations of Probability. Chicago: University of Chicago Press, 1950.

CARNAP, Rudolf. The Continuum of Inductive Methods. Chicago: University of Chicago Press, 1952.

CARNIELLI, Walter A.; EPSTEIN, Richard L. Pensamento crítico: o poder da lógica e da argumentação. São Paulo: Rideel, 2011.

CARTWRIGHT, Nancy. The Dappled World: A Study of the Boundaries of Science. Cambridge: Cambridge University Press, 1999.

CHANG, Hasok. Is Water H_2O? Evidence, Realism and Pluralism. Dordrecht: Springer, 2012.

CONDORCET, Marquis de. Essai sur l'application de l'analyse à la probabilité des décisions rendues à la pluralité des voix. Paris: Imprimerie Royale, 1785.

COVER, Thomas M.; THOMAS, Joy A. Elements of Information Theory. 2. ed. Hoboken: Wiley, 2006.

COX, Richard T. The Algebra of Probable Inference. Baltimore: Johns Hopkins University Press, 1961.

DASTON, Lorraine. Classical Probability in the Enlightenment. Princeton: Princeton University Press, 1988.

DASTON, Lorraine; GALISON, Peter. Objectivity. New York: Zone Books, 2007.

DAWID, A. Philip. The well-calibrated Bayesian. Journal of the American Statistical Association, v. 77, n. 379, p. 605-610, 1982. DOI: 10.1080/01621459.1982.10477856.

DE FINETTI, Bruno. Foresight: its logical laws, its subjective sources. In: KYBURG, Henry E.; SMOKLER, Howard E. (eds.). Studies in Subjective Probability. New York: Wiley, 1964. p. 93-158.

DE FINETTI, Bruno. Theory of Probability. Chichester: Wiley, 1974-1975. 2 v.

DESROSIÈRES, Alain. The Politics of Large Numbers: A History of Statistical Reasoning. Cambridge: Harvard University Press, 1998.

DURRETT, Rick. Probability: Theory and Examples. 5. ed. Cambridge: Cambridge University Press, 2019.

EAGLE, Antony (ed.). Philosophy of Probability: Contemporary Readings. New York: Routledge, 2010.

EINSTEIN, Albert. Investigations on the Theory of the Brownian Movement. New York: Dover, 1956.

FELLER, William. An Introduction to Probability Theory and Its Applications. New York: Wiley, 1950. v. 1.

FISHER, Ronald A. Statistical Methods and Scientific Inference. Edinburgh: Oliver & Boyd, 1956.

FRANKLIN, James. The Science of Conjecture: Evidence and Probability Before Pascal. Baltimore: Johns Hopkins University Press, 2001.

FUCHS, Christopher A.; MERMIN, N. David; SCHACK, Rüdiger. An introduction to QBism with an application to the locality of quantum mechanics. American Journal of Physics, College Park, v. 82, n. 8, p. 749-754, 2014. DOI: 10.1119/1.4874855.

GALAVOTTI, Maria Carla. Philosophical Introduction to Probability. Stanford: CSLI Publications, 2005.

GEBRU, Timnit et al. Datasheets for datasets. Communications of the ACM, New York, v. 64, n. 12, p. 86-92, 2021. DOI: 10.1145/3458723. Available at: https://arxiv.org/abs/1803.09010 Accessed: 3 Feb. 2026.

GELMAN, Andrew et al. Bayesian Data Analysis. 3. ed. Boca Raton: CRC Press, 2013.

GHIRARDI, GianCarlo; RIMINI, Alberto; WEBER, Tullio. Unified dynamics for microscopic and macroscopic systems. Physical Review D, College Park, v. 34, n. 2, p. 470-491, 1986. DOI: 10.1103/PhysRevD.34.470.

GILLIES, Donald. Philosophical Theories of Probability. London: Routledge, 2000.

GNEITING, Tilmann; RAFTERY, Adrian E. Strictly proper scoring rules, prediction, and estimation. Journal of the American Statistical Association, v. 102, n. 477, p. 359-378, 2007.

GOOD, I. J. The Estimation of Probabilities: An Essay on Modern Bayesian Methods. Cambridge: MIT Press, 1965.

GOODFELLOW, Ian; BENGIO, Yoshua; COURVILLE, Aaron. Deep Learning. Cambridge: MIT Press, 2016.

GUO, Chuan; PLEISS, Geoff; SUN, Yu; WEINBERGER, Kilian Q. On calibration of modern neural networks. In: Proceedings of the 34th International Conference on Machine Learning. 2017. p. 1321-1330.

HACKING, Ian. The Emergence of Probability: A Philosophical Study of Early Ideas About Probability, Induction and Statistical Inference. Cambridge: Cambridge University Press, 1975.

HÁJEK, Alan. Interpretations of probability. In: ZALTA, Edward N. (ed.). The Stanford Encyclopedia of Philosophy. Stanford: Metaphysics Research Lab, Stanford University, 2019.

HASTIE, Trevor; TIBSHIRANI, Robert; FRIEDMAN, Jerome. The Elements of Statistical Learning: Data Mining, Inference, and Prediction. 2. ed. New York: Springer, 2009.

HEISENBERG, Werner. Physics and Philosophy: The Revolution in Modern Science. New York: Harper, 1958.

HERNÁN, Miguel A.; ROBINS, James M. Causal Inference: What If. Boca Raton: Chapman & Hall/CRC, 2020.

HOEFER, Carl. Chance in the World: A Humean Guide to Objective Chance. Oxford: Oxford University Press, 2019.

HOWSON, Colin; URBACH, Peter. Scientific Reasoning: The Bayesian Approach. 2. ed. Chicago: Open Court, 2006.

HUMPHREYS, Paul. Why propensities cannot be probabilities. The Philosophical Review, v. 94, n. 4, p. 557-570, 1985.

HUYGENS, Christiaan. De ratiociniis in ludo aleae. Paris: 1657.

IMBENS, Guido W.; RUBIN, Donald B. Causal Inference for Statistics, Social, and Biomedical Sciences: An Introduction. Cambridge: Cambridge University Press, 2015.

JAYNES, Edwin T. Prior probabilities. IEEE Transactions on Systems Science and Cybernetics, v. 4, n. 3, p. 227-241, 1968. DOI: 10.1109/TSSC.1968.300117.

JAYNES, Edwin T. Probability Theory: The Logic of Science. Cambridge: Cambridge University Press, 2003.

JEFFREYS, Harold. Theory of Probability. 3. ed. Oxford: Oxford University Press, 1961.

JOYCE, James M. A nonpragmatic vindication of probabilism. Philosophy of Science, v. 65, n. 4, p. 575-603, 1998. DOI: 10.1086/392661.

KEYNES, John Maynard. A Treatise on Probability. London: Macmillan, 1921.

KNIGHT, Frank H. Risk, Uncertainty and Profit. Boston: Houghton Mifflin, 1921.

KOLMOGOROV, Andrei N. Foundations of the Theory of Probability. New York: Chelsea, 1956.

LAPLACE, Pierre-Simon. Essai philosophique sur les probabilités. Paris: Courcier, 1814.

LAPLACE, Pierre-Simon. Théorie analytique des probabilités. Paris: Courcier, 1812.

LEVI, Isaac. The Enterprise of Knowledge: An Essay on Knowledge, Credal Probability, and Chance. Cambridge: MIT Press, 1980.

LEWIS, David. A subjectivist's guide to objective chance. In: JEFFREY, Richard C. (ed.). Studies in Inductive Logic and Probability. Berkeley: University of California Press, 1980. v. 2, p. 263-293.

LEWIS, David. Humean supervenience debugged. Mind, v. 103, n. 412, p. 473-490, 1994. DOI: 10.1093/mind/103.412.473.

LIPTON, Zachary C. The mythos of model interpretability. arXiv:1606.03490, 2016. Available at: https://arxiv.org/abs/1606.03490 Accessed: 3 Feb. 2026.

MAHMOUD, Hosam M. Pólya Urn Models. Boca Raton: CRC Press, 2008.

MAXWELL, James Clerk. Theory of Heat. London: Longmans, Green, 1871.

MAYO, Deborah G. Error and the Growth of Experimental Knowledge. Chicago: University of Chicago Press, 1996.

MITCHELL, Margaret et al. Model cards for model reporting. arXiv:1810.03993, 2018. Available at: https://arxiv.org/abs/1810.03993 Accessed: 3 Feb. 2026.

MURPHY, Kevin P. Machine Learning: A Probabilistic Perspective. Cambridge: MIT Press, 2012.

NEYMAN, Jerzy; PEARSON, Egon S. On the problem of the most efficient tests of statistical hypotheses. Philosophical Transactions of the Royal Society of London. Series A, v. 231, p. 289-337, 1933.

OVADIA, Yaniv et al. Can you trust your model's uncertainty? evaluating predictive uncertainty under dataset shift. In: Advances in Neural Information Processing Systems. 2019. v. 32.

O'NEIL, Cathy. Weapons of math destruction: how big data increases inequality and threatens democracy. New York: Crown, 2016.

PEARL, Judea. Causality: Models, Reasoning, and Inference. 2. ed. Cambridge: Cambridge University Press, 2009.

PEARSON, Karl. The Grammar of Science. London: Walter Scott, 1892.

PETTIGREW, Richard. Accuracy and the Laws of Credence. Oxford: Oxford University Press, 2016. DOI: 10.1093/acprof:oso/9780198732716.001.0001.

PLANCK, Max. The Theory of Heat Radiation. Translated by Morton Masius. Philadelphia: P. Blakiston's Son & Co., 1914. Translation of the 2nd German edition of 1913.

POPPER, Karl R. A World of Propensities. Bristol: Thoemmes, 1990.

POPPER, Karl R. Realism and the Aim of Science. London: Hutchinson, 1983.

POPPER, Karl R. The propensity interpretation of probability. The British Journal for the Philosophy of Science, v. 10, n. 37, p. 25-42, 1959.

QUETELET, Adolphe. Sur l'homme et le développement de ses facultés, ou Essai de physique sociale. Paris: Bachelier, 1835.

QUINE, Willard V. O. Two dogmas of empiricism. The Philosophical Review, v. 60, n. 1, p. 20-43, 1951.

RAMSEY, Frank P. Truth and probability. In: BRAITHWAITE, R. B. (ed.). The Foundations of Mathematics and Other Logical Essays. London: Kegan Paul, 1931. p. 156-198.

REICHENBACH, Hans. Experience and Prediction: An Analysis of the Foundations and the Structure of Knowledge. Chicago: University of Chicago Press, 1938.

REICHENBACH, Hans. The Theory of Probability: An Inquiry into the Logical and Mathematical Foundations of the Calculus of Probability. Berkeley: University of California Press, 1949.

ROTHMAN, Kenneth J.; GREENLAND, Sander; LASH, Timothy L. Modern Epidemiology. 3. ed. Philadelphia: Lippincott Williams & Wilkins, 2008.

RUBIN, Donald B. Estimating causal effects of treatments in randomized and nonrandomized studies. Journal of Educational Psychology, v. 66, n. 5, p. 688-701, 1974.

RYLE, Gilbert. The concept of mind. London: Hutchinson's University Library, 1949.

SALMON, Wesley C. Scientific Explanation and the Causal Structure of the World. Princeton: Princeton University Press, 1984.

SAVAGE, Leonard J. The Foundations of Statistics. New York: Wiley, 1954.

SEIDENFELD, Teddy. Why I am not an objective Bayesian. Theory and Decision, v. 11, p. 413-440, 1979.

SELLARS, Wilfrid. Empiricism and the philosophy of mind. In: FEIGL, Herbert; SCRIVEN, Michael (eds.). Minnesota studies in the philosophy of science. Minneapolis: University of Minnesota Press, 1956. v. 1, p. 253-329.

SHAFER, Glenn. A Mathematical Theory of Evidence. Princeton: Princeton University Press, 1976.

SKYRMS, Brian. The Dynamics of Rational Deliberation. Cambridge: Harvard University Press, 1990.

SOBER, Elliott. Evidence and Evolution: The Logic Behind the Science. Cambridge: Cambridge University Press, 2008.

SPIRTES, Peter; GLYMOUR, Clark; SCHEINES, Richard. Causation, Prediction, and Search. 2. ed. Cambridge: MIT Press, 2000.

STIGLER, Stephen M. The History of Statistics: The Measurement of Uncertainty Before 1900. Cambridge: Harvard University Press, 1986.

SUPPES, Patrick. Representation and Invariance of Scientific Structures. Stanford: CSLI Publications, 2002.

TALEB, Nassim Nicholas. The Black Swan: The Impact of the Highly Improbable. New York: Random House, 2007.

VAN FRAASSEN, Bas C. Laws and Symmetry. Oxford: Oxford University Press, 1989.

VENN, John. The Logic of Chance: An Essay on the Foundations and Province of the Theory of Probability. London: Macmillan, 1866.

VON MISES, Richard. Probability, Statistics and Truth. 2. ed. New York: Dover, 1957.

VON PLATO, Jan. Creating Modern Probability: Its Mathematics, Physics and Philosophy in Historical Perspective. Cambridge: Cambridge University Press, 1994.

WALLEY, Peter. Statistical Reasoning with Imprecise Probabilities. London: Chapman and Hall, 1991.

WILLIAMSON, Jon. In Defense of Objective Bayesianism. Oxford: Oxford University Press, 2010.

WITTGENSTEIN, Ludwig. Philosophical investigations. 4. ed. Oxford: Wiley-Blackwell, 2009.